THE SCIENTIFIC PHOTOGRAPHER

PLATE I

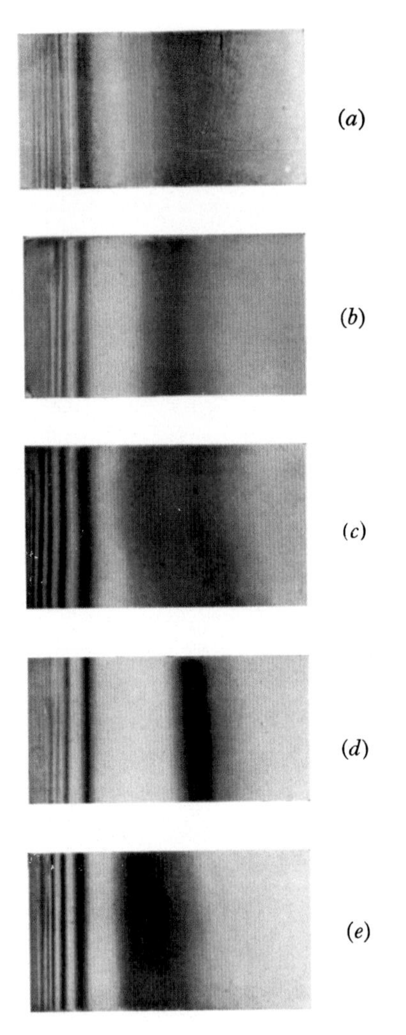

Fig. 11. Reproduction of colours of soap film.

THE SCIENTIFIC PHOTOGRAPHER

BY

A. S. C. LAWRENCE, Ph.D.

CAMBRIDGE
AT THE UNIVERSITY PRESS
1941

CAMBRIDGE
UNIVERSITY PRESS

University Printing House, Cambridge CB2 8BS, United Kingdom

Published in the United States of America by Cambridge University Press, New York

Cambridge University Press is part of the University of Cambridge.

It furthers the University's mission by disseminating knowledge in the pursuit of education, learning and research at the highest international levels of excellence.

www.cambridge.org
Information on this title: www.cambridge.org/9781107698581

First published 1941
First paperback edition 2014

A catalogue record for this publication is available from the British Library

ISBN 978-1-107-69858-1 Paperback

ERRATA

p. 71, *line* 7, *for* meter *read* shutter
p. 74, *line* 3, *last word, for* violet *read* orange
p. 97, *lines* 2 *and* 3, *for* (*a*) *read* (*b*) *and for* (*b*) *read* (*a*)

CONTENTS

PLATES IN COLOUR

BIBLIOGRAPHY

EDER. *Ausführliches Handbuch der Photographie.* 1920–30. (In German.)

HAY, A. and VON ROHR, M. *Handbuch der wissenschaftlichen und angewandten Photographie.*

CLERC, L. P. *Photography.* (English edition edited by G. E. Brown, 1937.)

MACK, J. E. and MARTIN, M. J. *The Photographic Process.* 1939.

NEBLETTE, C. B. *Photography.* 1937.

SPENCER, D. A. *Colour Photography in Practice.* 1938.

CLARK, W. *Photography by Infra-Red.* 1939.

VON ROHR, M. *Geometrical Investigation of the Formation of Images in Optical Instruments.* Translated by R. Kanthack. 1920.

GLEICHEN, A. *Theory of Modern Optical Instruments.* Translated by H. H. Emsley and W. Swaine. 1921.

CONRADY, A. E. *Applied Optics and Optical Design.* 1929.

BARNARD, J. E. and WELCH, F. V. *Practical Photo-Micrography.* 1936.

MORGAN, W. D. *Synchroflash Photography.* 1939.

OTTLEY, D. C. *The Cinema in Education.* 1939.

LESTER, H. M. *The Photo-Lab Index.*

MORGAN, W. D. and LESTER, H. M. *Graphic and Graflex Photography* (1940). Although this book deals with the Kodak reflex it should be of value to all users of this type of camera.

Catalogue of the Epstean Collection. University of Columbia Press, 1937. This is the finest collection of photographic works in existence. The catalogue is therefore the most complete bibliography of photographic bibliography.

Numerous trade pamphlets containing valuable information are supplied by the more important manufacturers of cameras and photographic materials. These are free or sold for a few pence.

The Ilford Manual of Photography is one of the best books of general instructions for elementary workers.

The British Journal Photographic Almanac, which appears annually, is a useful source of formulae. Its advertisement section is a valuable survey of the latest apparatus and materials on the market.

Photo-Technique. A technical photographic monthly published by the McGraw Hill Publishing Co.

PREFACE

During the last twenty years or so very great advances have been made in photography, both in the design of apparatus and in the performance of materials. It is doubtful, however, whether there has been a commensurate advance in the quality of work turned out. Spectacular results have been obtained, but these are new applications which would not have been possible earlier and they have no bearing upon the general level of photography. In scientific work, results of very great value have been obtained by using the photographic plate as a detector of new radiations, but the photographic technique involved is rudimentary. Full use of photography is certainly not made in scientific laboratories. Nor has any organized attempt been made to use cinematography either for research or for serious teaching.

The reasons for this failure are complex, but all boil down to one—ignorance. Ignorance of the possibilities and limitations of apparatus and materials. The subject is now too complex to be 'picked up' from a few demonstrations in the dark room. Previous experience as a 'button-pusher' may implant ideas which are more harmful than complete ignorance. Far too many laboratories possess no photographic equipment and no member of the staff with real technical knowledge. In these circumstances, photography is used only when there happens to be working there an amateur whose enthusiasm can be exploited when necessary. The parochial spirit of Departments is a barrier against any communal service by a skilled worker. Consequently, considerable sums are spent on work done by outside commercial firms.

It would be quite easy to produce a collection of striking scientific photographs, but such a work would be of little value, since the chosen examples would be striking because of the scientific value or the nature of the subjects rather than because of any special technique. In any case, no general information of photography as a tool is given in such a book and this is just what is wanted by the scientific worker. This book attempts to give a short account of the possibilities and limitations of photography—primarily for the scientific worker but also for the serious amateur who has an elementary knowledge of chemistry and physics. For there is no special scientific photography for scientists. All the branches of the subject used by amateurs and professionals are used at some time in scientific work. The book has been kept short by omitting detailed instructions of the sort included with materials when bought and by giving basic information rather than detailed minor variations.

A serious limitation to the use of photography in publications is the high cost of printing. This is particularly serious in the case of scientific journals, but there is no reason why lantern slides should not be used more. The bogey of cost is supposed to explain the low level of illustration of most scientific books, but occasionally a book appears which suggests that this bogey is not so real. An excellent example is *Animals without backbones*, by Ralph Buchsbaum (371 pp.), published by the University of Chicago Press at 18*s*. 6*d*. (1938).

I am indebted to Mr G. W. W. Stevens and to Mr P. R. P. Claridge for reading the manuscript and for useful suggestions. From Messrs Kodak, Ilford, Zeiss Ikon and Agfa I have received much valuable information—more, in fact, than I have been able to use in detail. For illustrations, I am indebted to the following: Messrs Kodak (1*a*, 9, 47, 48, 56, Messrs Ilford (8 *b*, 57, 66), Messrs Ilford and *The Times* (79), Messrs Dufay-Chromex, Ltd. (45, 46), Messrs Zeiss Ikon (40, 72), Prof. Svedberg (1 *b*), Mr Webb, 2, Mr G. A. Jones, 15, Messrs Ensign, 33, Messrs Leica, 41, Dr H. Dunlop, 69, Major G. W. G. Allen, 70, Mr W. E. Woolfe and Science Films, Ltd., 71, Dr W. T. Astbury, 80, Prof. G. I. Finch, 81. The remaining illustrations are the author's.

A. S. C. L.

4 *September* 1940

CHAPTER I

THE BASES OF PHOTOGRAPHY

Sensitivity of silver compounds to light and the discovery of photography

It has been known for a very long time that certain substances are sensitive to light. In particular, silver compounds are darkened when exposed to sunlight. Wedgwood, son of the famous potter, and Sir Humphry Davy made a number of observations on silver chloride. They succeeded in making prints, but failed to make these permanent. Meanwhile Daguerre had made, accidentally, a discovery of fundamental importance. His light-sensitive material was silver iodide, which he prepared by the action of iodine vapour on sheets of silver. Light alone produces metallic silver after very long exposure, but Daguerre found that a plate, exposed for a time insufficient to show a visible image, when placed in a cupboard which contained some mercury, produced a visible image. He had discovered the principle of development and, incidentally, of the latent image. This principle is the basis of all subsequent photography.[1] The light-sensitive system is exposed to light for a short time. Inspection shows no obvious change. When 'developed' an image of silver is formed on the invisible latent image. The second major discovery was that of 'fixing'. The earlier workers had obtained easily some sort of image, but they were not able to remove the unchanged light-sensitive substance and their prints were therefore not permanent but blackened slowly all over.

The direct forerunner of modern photography was Fox Talbot. In 1837 he obtained photographs using a camera obscura with a sheet of paper washed with silver nitrate as his sensitive material. These he rendered permanent by washing in a solution of common salt. Herschel had discovered hypo, sodium thiosulphate, $Na_2S_2O_3$, in 1819 and observed its extreme efficiency in dissolving silver halides.[2] Earlier in 1837, the Rev. J. B. Reade had used paper

[1] Photographs can be taken using thallium iodide and bromide in place of silver salts. (Lüppo-Cramer, *Die grundlagen der photographischen negativverfahren*, p. 584 and W. J. G. Farrer, *Photogr. J.* 1936, **76**, p. 486.)

[2] Herschel appears to have been the originator of the word 'photography'.

soaked in silver nitrate and washed over with a solution of gallic acid. This greatly increased the sensitivity, and he used the process to take photomicrographs, using the sun as light source. He also seems to have been the first person to make use of hypo as a fixer.

The subsequent history of photography is fundamentally no more than an improvement of Fox Talbot's and Daguerre's and Reade's early methods. The first stage was the use of glass plates coated with a suitable mixture of silver salt and some material to hold it in a fine state of dispersion. The first material used for this purpose was collodion. This had the serious drawback that the plates had to be prepared immediately before using. They were coated with a layer of collodion containing a mixture of soluble bromides and iodides. As soon as the collodion had set it was bathed in a solution of silver nitrate. Silver halides were then formed in the collodion, which also retained the other decomposition products of the reaction. The plate had to be used wet, otherwise crystallization of these products occurred in the emulsion. Very fine results were obtained nevertheless with these plates, and collodion plates are still unrivalled where extremely fine grain is required. It should be noted also that development was 'physical' from Fox Talbot's discovery of development of the silver image up to collodion wet plates. The source of silver was the excess of silver nitrate remaining in the negative, and this was reduced by a ferrous salt. In the earliest dry plates, from which the necessary supply of silver nitrate had been washed out, silver nitrate had to be added to the developing solution. This need for physical development must be considered in connection with the remarkable sharpness of the early collodion wet-plate negatives.

The first big improvement consisted in the use of gelatine as the support for the silver halide. The soluble salts, formed by the decomposition of the silver nitrate and soluble halides used, were washed out from the suspension of silver halide in gelatine, which received the name of 'emulsion'. The name is, of course, entirely incorrect, since the system is not an 'emulsion', but the name has persisted.[1] These plates, since they contain only silver halide and

[1] The name 'emulsion' belongs only to stable colloidal dispersions of one liquid in another. Photographic 'emulsions' are suspensions of silver halide microcrystals in the gelatine support.

gelatine, can be dried without crystallization taking place. It was also found that gelatine plates were intrinsically more sensitive to light than the collodion ones. The gelatine appears to have some specific effect upon the silver halide, which improves it in this manner. This point will be discussed later (p. 9). The gelatine prevents immediate precipitation of coarse amorphous silver chloride so that a fine crystalline form is reached. Fig. 1 (*a*) shows

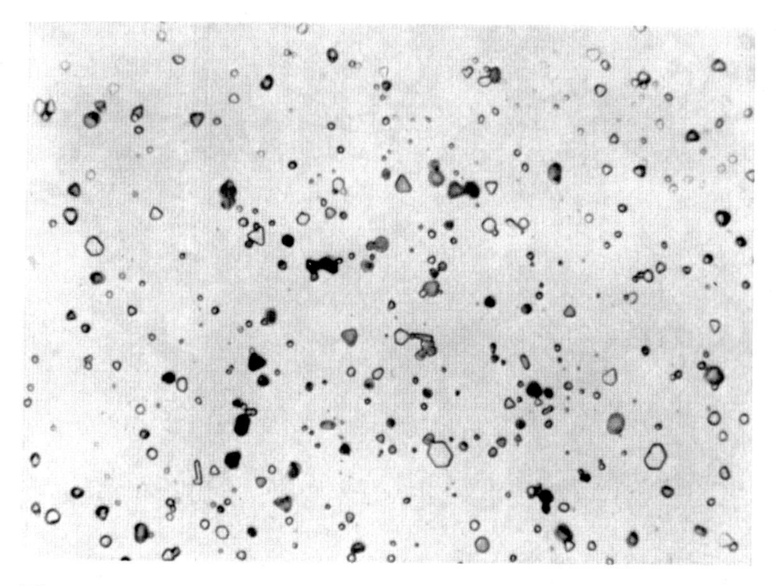

Fig. 1 (*a*). Grains of silver bromide in photographic emulsion. × 1500.

the appearance of the silver halide in an emulsion magnified 1500 times.

The next big advance was the use of celluloid as support for the emulsion in place of glass. This led to the roll film, film pack, and cut film. These have a number of advantages over plates. For convenience of handling, the roll film is superior, and any type of gelatine film has the great advantage for storage that it is very much lighter and takes up less space. Loss of definition by halation is reduced (p. 27).

The subsequent improvements in photographic emulsions consisted of increasing speed, fineness of grain, and sensitivity to all parts of the visible spectrum. In recent years their sensitivity has

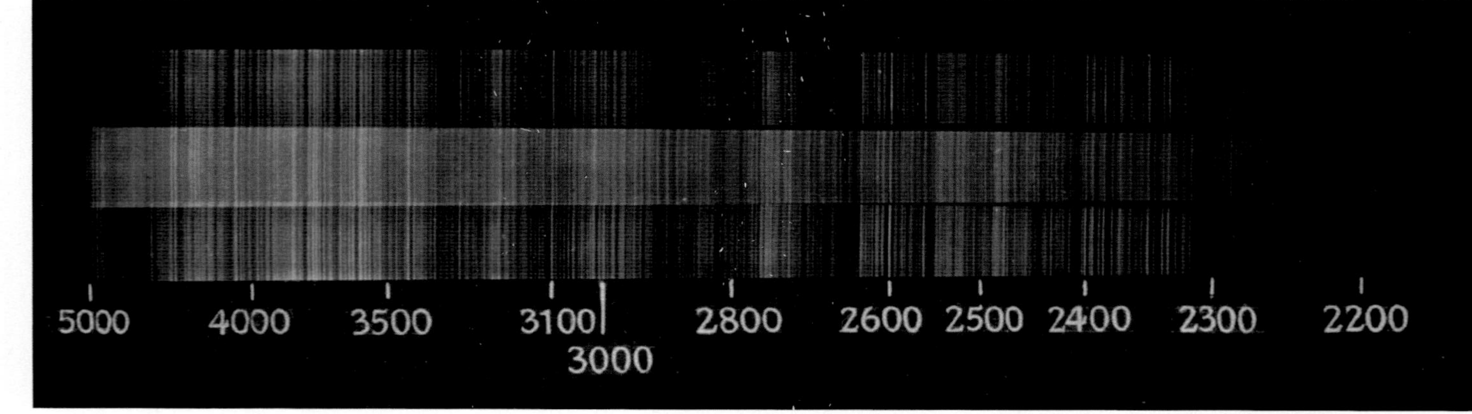

Fig. 2. Spectrogram of iron arc.

been extended far into the infra-red and ultra-violet regions.[1] It is also found that emulsions prepared for use with visible light are sensitive to numerous radiations of short wave-length, such as X-rays and electrons, and also to α-particles. Becquerel (1896) discovered radio-activity photographically.

Emulsions will be discussed first since, in some cases, they are the only photographic material used in scientific work. The object of photography in scientific work is almost invariably to obtain the most accurate possible record. This includes two and sometimes three quite distinct points: (1) accuracy of reproduction and separation of details—generally called resolving power; (2) accuracy of recording tone values, e.g. recording the range of luminosities on a scale which is identical with, or which bears a simple relation to, the original; (3) equivalence of recording of different colours on the scale of perception of the human eye: or the exact opposite—increasing colour contrast to show important detail more clearly against unimportant confusing background.

Fig. 2 is a spectrogram which illustrates all three points. The large range of wave-lengths from an iron arc is recorded approximately uniformly and the closely distributed lines are separated distinctly. The method used to obtain images of approximately equal intensity of wave-lengths of very widely different actinic action was that described on p. 119.

Theory of photographic action

At one time there was an idea that photographic action could be explained by a single theory. Now, however, it is clear that we have at least three more or less independent points to explain:

(1) The nature of the emulsion and the cause of its sensitivity to light,

(2) The mechanism of the absorption of energy and formation of the latent image,

(3) The nature of development.

The appearance of the emulsion suggests that it is a suspension of crystalline silver halide (Fig. 1 *b*). The form is flat hexagonal

[1] Ultra-violet radiations were first detected by their active effect on silver chloride (Wollaston, 1804). Scheele in 1777 had observed that the blue violet end of the spectrum had greatest actinic action on silver chloride.

plates (silver chloride and bromide belong to the cubic system), but the boundaries are somewhat fuzzy. The fact that the silver halide has the external appearance of a crystalline substance does not mean very much. In the first place, we know that crystals are not to be regarded as homogeneous and that their most reactive parts are edges, corners and cracks. Now although the sensitivity of a silver-halide emulsion is increased by processes which are accompanied by increase of size of the crystallites, these processes

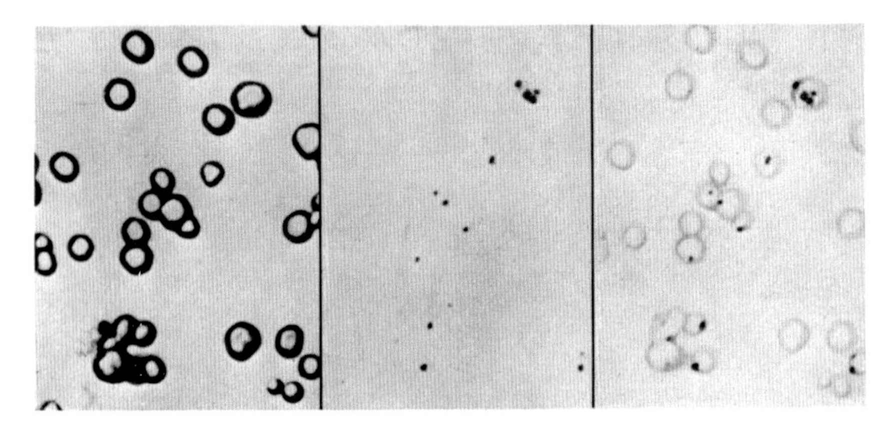

Fig. 1 (*b*). Svedberg development nuclei.

also involve formation of silver and silver oxide or sulphide; and examination under the highest magnifications shows after suitable treatment that there are in the hexagonal plates black specks[1] which act as nuclei for subsequent development (fig. 1 (*b*)).

The brilliant work of Sheppard has shown that complex organic substances containing sulphur are present in gelatine and that, in the processes of ripening the emulsion, silver sulphide specks are formed (p. 9).

Recently, Gurney and Mott[2] have attacked the problem of absorption of energy and have shown that a dispersion of silver or silver-sulphide specks in a silver-halide lattice provides a very efficient electron trap. An electron liberated by the action of light on the emulsion is free to move through the silver-halide lattice

[1] Svedberg, *Photogr. J.* 1922, **62**, pp. 186, 310.
[2] R. W. Gurney and N. F. Mott, *Proc. Roy. Soc.* A, 1938, **164**, p. 151.

until it is trapped by a particle of silver. If the work function of the metal is greater than the energy of the electron, then the latter is in a state of lower potential energy when trapped by a metal particle. We know that, once a particle of metal exists in a crystallite, development will convert the whole of the crystallite into metallic silver, given sufficient time. The silver or silver-sulphide specks already described provide the means of trapping electrons. Mott's[1] explanation of development is as follows: When a molecule of developer comes into contact with a silver-halide crystal, it cannot lose an electron to it because the energy level of the electrons in the adsorbed layer of developer molecules lies between the highest occupied level of the metal and the lowest level of the conduction band of the silver halide. But every time that a molecule comes into contact with an uncharged particle of silver, it will lose an electron and the silver will gain a negative charge. It will then attract to itself silver ions.

The nature of photographic emulsions

Photographic 'emulsions' are not emulsions. They are suspensions of silver halide in gelatine. In their earlier stages of preparation the silver halide is formed by double decomposition of silver nitrate and soluble halide in aqueous solution in the presence of gelatine. It is then a protected colloidal dispersion of silver halide. It is allowed to set to a hard gel, cut up and the water-soluble products of the reaction washed out. The emulsion is then heated with water and 'ripened'. 'Ripening' is the trade name given to the two separate processes which increase the sensitivity of emulsions to light—particle growth and formation of nuclei. First, particle growth occurs until the 'emulsion' is a suspension of micro-crystals of silver halide in gelatine. How far we must regard the action of the gelatine as a purely mechanical one of suspending and insulating the micro-crystals and how far there is still definite interaction between the gelatine and the surfaces of the crystals is not clear. Their diffuse outlines seen under the microscope suggest the latter. Silver chloride belongs to the cubic system, but the crystals in an emulsion always have the characteristic form of thin hexagonal plates shown in fig. 1 (*a*). Their size is very important since it seems that, other conditions being the same, the largest

[1] N. F. Mott, *Photogr. J.* 1938, **78**, p. 286.

crystals are most sensitive to light. Commercial emulsions are remarkably uniform in particle size. Fig. 3 shows the distribution of size in an emulsion.

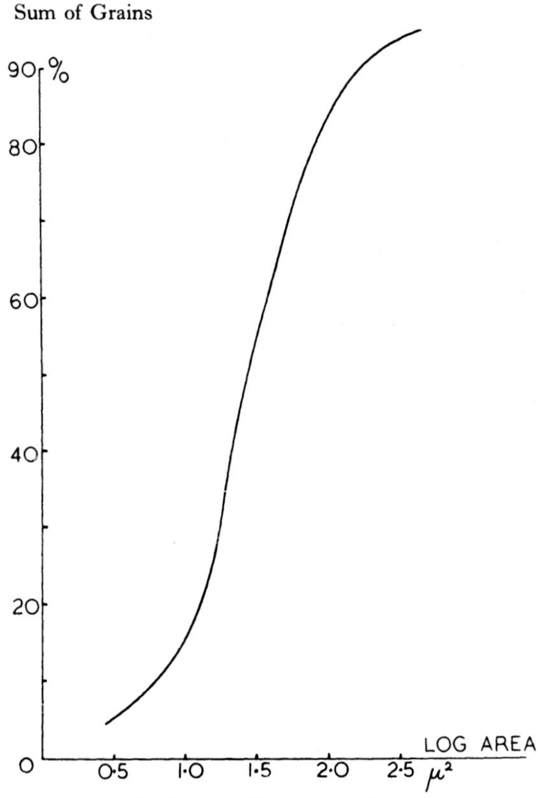

Fig. 3. Grain-size distribution in emulsion.

Abney's observation that gelatine or other colloid is not essential in the first stage of the preparation of the emulsion, since glycerine can replace them as the protector against coagulation of the precipitate, also suggests that there is interaction between the protecting substance and the silver halide amounting to formation of a surface complex.

The process of ripening is sometimes visualized as a simple 'ageing' or particle growth analogous to the thermodynamically simple case of growth of large particles of a liquid at the expense of smaller ones which have a higher vapour pressure. Ripening is, however, much more complex. In the first place, the micro-crystals are not of uniform energy over their surfaces, the corners and edges being most reactive. Ripening consists of two stages, only the first of which is accompanied by grain growth.[1] This first 'ageing' is carried out by heating in the presence of excess of potassium bromide and consists of an increase of crystal size. The potassium bromide is then washed out and the second stage takes place, sometimes in the presence of ammonia. It is believed that the extra sensitivity resulting from this second stage is due to formation of sensitive nuclei either of silver or silver sulphide in the micro-crystals and not to further crystal growth.

Following observations that the final sensitivity of emulsions varied with the history and source of the gelatine used, Sheppard, in a brilliant and classical research,[2] showed that this was due to the presence of organic sulphur compounds in the gelatine in varying amount. During the second stage of the ripening, these break down and silver-sulphide specks are formed. The manner in which these increase the speed of the emulsion has already been described by the theory of Gurney and Mott.

The exact nature of the organic sulphur compound in the gelatine is uncertain owing to its extremely small amount, which is from one to three or four parts in a million. It is analogous to allyl isothiocyanate. This reacts with ammonia to form a thio-carbamide:

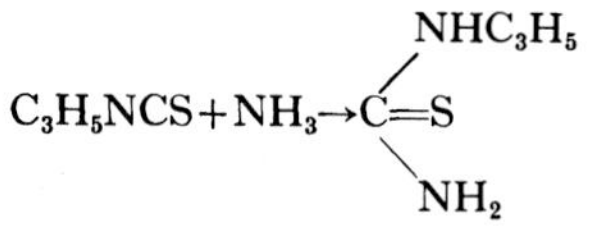

$$C_3H_5NCS + NH_3 \rightarrow C(=S)(NHC_3H_5)(NH_2)$$

and it is the group (probably acting in a tautomeric form)

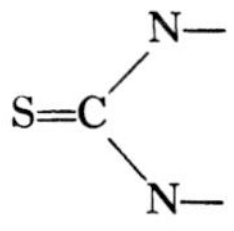

$$S{=}C\begin{matrix} \diagup N- \\ \diagdown N- \end{matrix}$$

which is responsible for sensitizing.

[1] E. J. W. Verwey (*Proc. K. Akad. Wet. Amst.* 1933, **36**, 2, pp. 225–33) has found that the ageing of aqueous sols of silver iodide falls into two parts. The first is very rapid and is accompanied by marked decrease of amount of adsorbed ions. This is followed by a slower ageing consisting of increase of grain size. He suggests that the first is an improvement of an originally very irregular crystal habit accompanied by a decrease of actively adsorbing corners and edges, while the second is a normal crystal growth.

[2] *Photogr. J.* 1925, **65**, p. 380 and 1926, **66**, p. 399.

Sensitivity of emulsions and the characteristic curve

In 1890 Hurter and Driffield made a systematic examination of the response of emulsions to light and devised a system of speed rating which has survived, with added complications, to this day. They determined the density of the image produced by exposures of varying length followed by standard development. When these densities are plotted against the log of the exposure an S-shaped curve is obtained (fig. 4). The reason for using the log of the

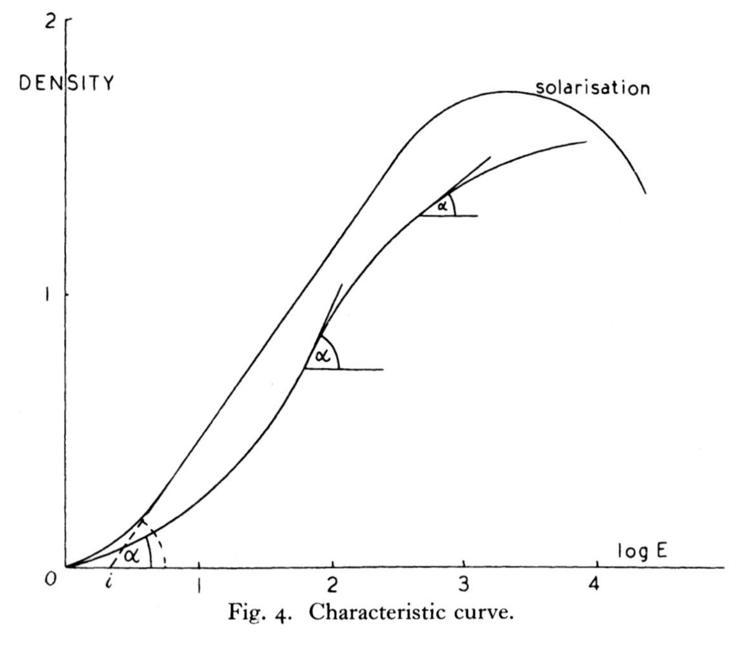

Fig. 4. Characteristic curve.

exposure is because the absorption of light by the developed image is related to the amount of silver present according to the Lambert-Beer law, which states that

$$I = I_0 e^{-kc\theta},$$

where I is the amount of light transmitted through a depth θ, I_0 is the intensity of incident light, k is the molecular absorption coefficient, and c is the concentration of absorbing substance. Taking logs, we get

$$k = \frac{\log_e I_0 - \log_e I}{c\theta}.$$

When logs to the base 10 are used, k is the absorption coefficient and, for photographic purposes, the density. Density is defined as log opacity, which is defined as I_0/I.

The characteristic curve of an emulsion is peculiar in that, for low exposures, the response is disproportionately low; and for very long exposures, the linear relation also breaks down again and finally the density becomes less. This last phenomenon is called 'solarization'. It has little bearing on photography, since the exposures required are of the order 10^5 times that at the point i. Hurter and Driffield estimated speed by extending the linear portion of the characteristic curve to intercept the exposure axis. The value of this intercept was called by them the 'inertia'. Their H. and D. speed numbers are obtained by dividing 34 by the value of the inertia, expressed in candle/metre/seconds.

The slope of the linear part of the curve gives the density gradient and is always denoted by the Greek letter γ and called the gamma of the emulsion.

From a characteristic curve we can get a number of values which may be useful:

(1) Maximum density.

(2) Contrast gradient at any part of the curve.

(3) Contrast gradient over the straight line portion, i.e. γ.

(4) Latitude, i.e. range of light intensities recorded by the straight line portion. That is, the length of the projection, measured in exposure units, of the straight line part of the characteristic curve upon the log E axis.

(5) Scale, i.e. range of intensities reproducible as density differences.

(6) i, the inertia.

(7) The angle α (fig. 4).

It should be noted that the value of i, the inertia, will vary with time of development. Developers containing bromide commonly give characteristic curves which intersect at a point below the log E axis (fig. 63, p. 121). Unless stated to the contrary, characteristics such as i, which vary with time of development, are assumed to have been observed when development had been taken to gamma unity.

Hurter and Driffield's method of comparing speeds of emulsions was satisfactory so long as they all had characteristic curves of the same shape. This no longer holds good (see fig. 5). We now need to compare curves of different shape and no longer the slope of a line. The Scheiner system states speed by measuring the smallest amount of light which is required to give an image on development

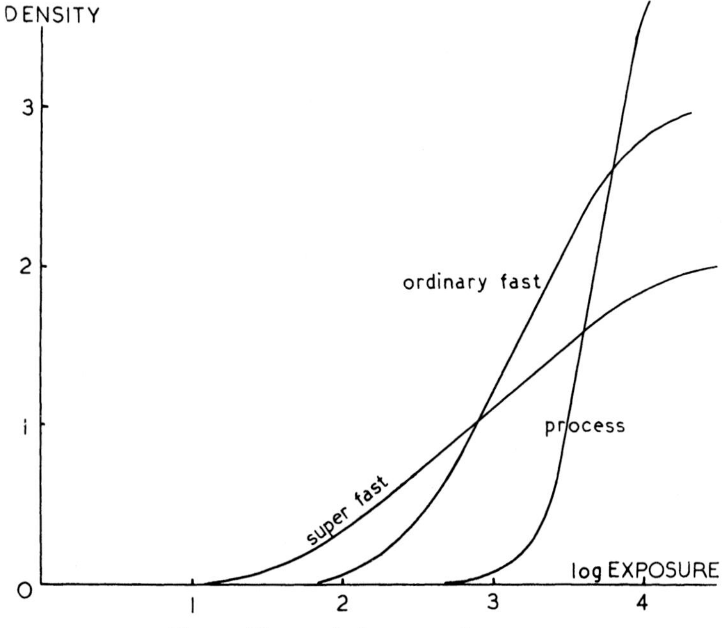

Fig. 5. Types of characteristic curves.

which can just be differentiated from the unexposed emulsion. This method obviously gives useful information concerning the speed for especially short exposures. The Din system exposes an emulsion to a standard source of light through a step wedge in which the steps correspond with an increase of density of 0·1. The emulsion is then subjected to prolonged development and the speed stated as the density step in the wedge beneath which the density of the image is 0·1 above fog, the latter not to exceed 0·4. The chief objection to these systems is that the methods of estimation bear no simple relation to normal working conditions owing to the serious over-development in the one case and under-exposure in the other. Fig. 6 shows the approximate relation between the Scheiner, Din and H. and D. speed ratings.[1]

For accurate reproduction of tone values, it is necessary to make use of the linear part of the characteristic curve. Further, for each emulsion, there is an infinite number of characteristic curves according to the development time. It is therefore necessary to choose a development time (and, of course, concentration and

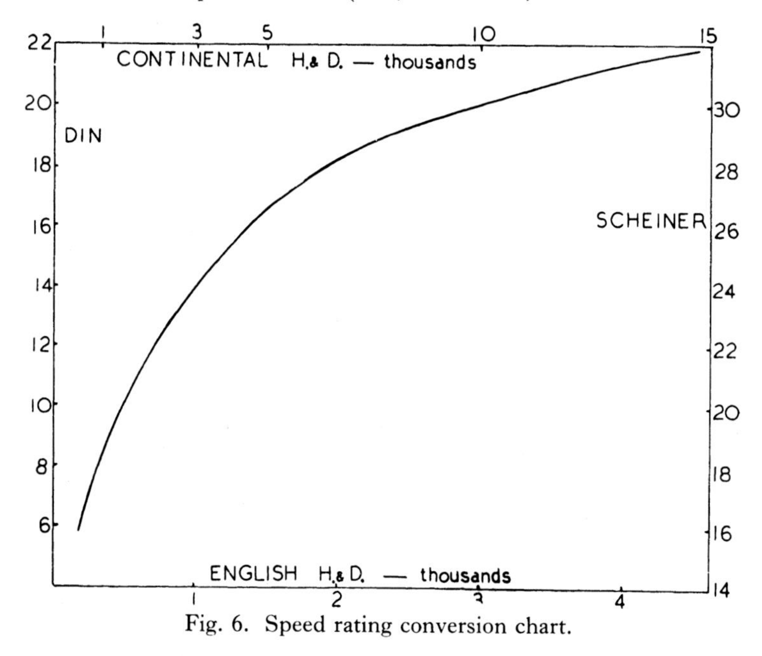

Fig. 6. Speed rating conversion chart.

temperature) which will give a predetermined gamma. This gamma is not to be chosen arbitrarily but so that it will fit the gamma of the paper on which the printing is to be done. It is common to meet photographers who consider seriously the gamma of their negatives without any attention to the characteristics of the paper that they propose to use.

Much has been written on the theory of correct tone reproduction, but most of it is of little theoretical value and less photographic. In the majority of subjects the range of brightness between high-lights and deepest shadows is considerably greater than

[1] Continental H. and D. numbers are stated arbitrarily as about three times the English ones.

that which can be realized by any paper, since the maximum brightness of paper is limited, nor can true dead black be obtained, so that the range of brightness in a print must be less than that in the original subject. Now it is obviously possible to make prints in which the tones are rendered correctly relative to one another but on a scale different from the original and necessarily flatter. When brightness is wanted in the print, as it usually is, this flattening is bad; and it is very much better to decide which part of the tones are of least importance and to sacrifice them.

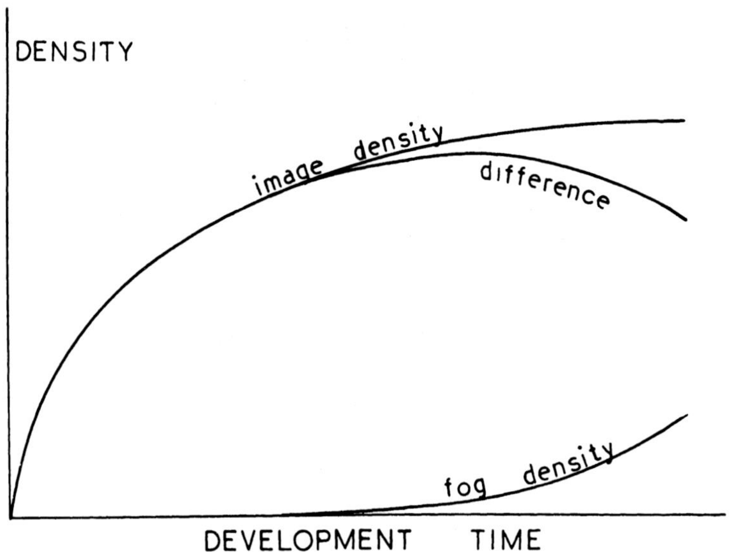

Fig. 7. Effect of fog on negative density.

This problem of obtaining correct tones (or any desired degree of incorrectness) will be discussed more fully, with examples, in Chapter v. Meanwhile it must be recognized that development time must be such that the negative will print on the paper chosen. The best method is to make negatives fit medium-grade paper; then using soft or contrasty papers allows latitude if it is found better to alter the gradation of tones. Some workers with miniature cameras systematically keep their development time short so that their 'normal' negatives require contrasty paper to print. This short development time keeps grain small, but the method has the obvious defect that contrast cannot be increased by using a more contrasty paper. It cannot be too strongly emphasized that development time is not just the time that the negative requires for all the desired detail to appear. Extending the time does bring up under-exposed objects and shadow detail but only at the cost of blocking up the high-lights so that all the light tones appear on the print as whites. Errors of exposure cannot be corrected by altering the development time.

It must be remembered that, while the density of an image is building up, general fog is also increasing. Fig. 7 shows the rates of growth of the two. In considering a characteristic curve the H. and D. method is quite satisfactory when comparing a family of curves of similar shape. As the examples in fig. 5 show, the characteristic curves of modern high-speed emulsions have a variety of shapes. Considerable difficulty, therefore, arises from the increasing size of the foot of the curve. The lowest point on the curve to be used in determining the speed of the film has been and still is a matter for difference of opinion. It may be taken as the point at which the density has reached some definite point value above fog; the choice of this value is more or less arbitrary. Or it may be chosen as the value at which gamma has reached some chosen value. Equally, the highest point may be defined as the point of maximum density or some point slightly below this. Equally, some completely arbitrary value of density may be chosen. There is some theoretical justification for this since the upper part of the characteristic curve is not used in practice. The top point may also be chosen as some definite small value of gamma.

An under-exposed negative is one in which the shadows are disproportionately dark.[1] An over-exposed negative has its disproportion in the high-lights. In each case that particular tone scale is overcrowded. In the print the shadows will be blocked up in under-exposure and the high-lights too blank in over-exposure. **These faults cannot be remedied by varying the exposure during printing.**

The characteristic curve shows also that to get a linear increase of density requires a logarithmic increase of exposure. Always

[1] In discussing negatives it is usual to describe them according to the sort of positive print they will give. Thus shadows and high-lights refer to these parts of the positive.

therefore in making a series of exposures, when the correct one is unknown, use a logarithmic scale, e.g. 2, 4, 8, 16, 32 times and so on, and not 2, 3, 4, 5, 6 times.

Colour sensitivity of emulsions

The response of an ordinary[1] emulsion to white light is shown in fig. 8 (*a*), curve 1. It is seen that a plate or film of this sort is blind to everything except the violet and blue end of the spectrum. In addition, even ordinary emulsions are very sensitive to the ultra-violet. The human eye, on the other hand, is most sensitive to the yellow.[2] In fig. 8 (*a*), curve 2 is for an orthochromatic and curve 3

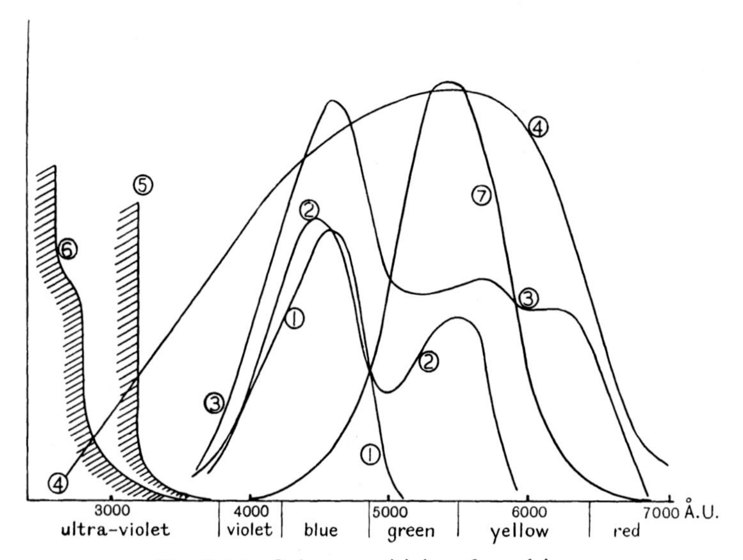

Fig. 8 (*a*). Colour sensitivity of emulsions.

for a panchromatic emulsion (see below). Curves 5 and 6 show the limits of transmission of light by glass and gelatine respectively. Curve 4 shows the response of a photo-electric cell and curve 7 that of the human eye.

It has been found that addition of certain dyes to the emulsion

[1] 'Ordinary' is used to describe emulsions not specially sensitized to the green, yellow and red part of the spectrum.

[2] In very weak illumination, the maximum sensitivity of the human eye travels towards the green-blue, 5300 A.U. This is called the Purkinje effect and is very useful in dark-room practice since this region is also the least actinic.

before coating, or by bathing the plate or film in a dilute solution of the dye, renders it sensitive to other parts of the spectrum.[1] In general we may say that this extra sensitivity to colour is restricted to those colours which the dye absorbs. Thus a dye which sensitizes for green and yellow will be of violet colour. It has been shown that the amount of dye required to produce the maximum sensitizing effect is strictly proportional to the area of silver halide, and it is suggested that the dye is present as a monomolecular

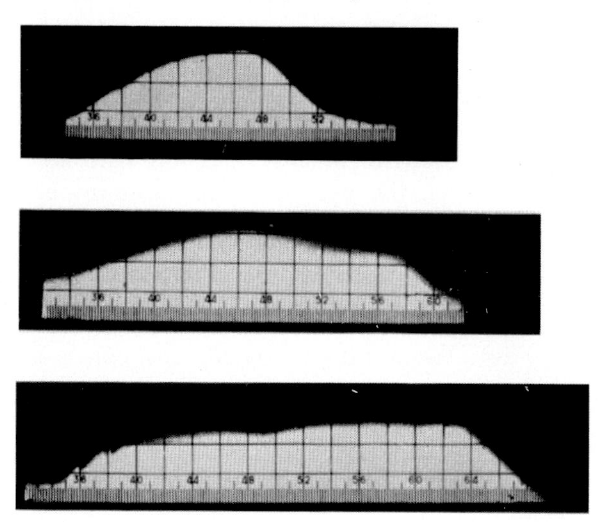

Fig. 8 (*b*). Spectrograms of colour sensitivity.

layer.[2] More dye than the optimum amount decreases sensitivity by absorbing light. Absorption spectra of dyes on silver halide are of the same form as those of the solutions of the dyes in inorganic salt solutions. There is some complexity about the matter, since many of these complex dyes are differently coloured according to their state of dispersion. They may be in solution as ions, molecules or colloidal micelles, each having its own absorption spectrum.[3] The first colour-sensitive emulsions prepared were sensitive to green and yellow but not red and were called *orthochromatic* (fig. 8 (*a*), curve 2). Since then, *panchromatic* emulsions have been

[1] Vogel, 1873 (using collodion emulsions).
[2] *J. Chem. Phys.* 1937, **7**, p. 878 *et seq.*
[3] Jelley, *Nature*, 1937, **139**, p. 631.

made which are sensitive to the whole spectrum although their sensitivity curve is still far from that of the human eye (fig. 8 (*a*), curves 3 and 4). In particular their sensitivity is still much greater in the violet and blue end of the spectrum, but this extra sensitivity is readily corrected by the use of a yellow-green filter in front of the lens. Fig. 8 (*b*) shows the sensitivity distribution as spectrograms. Emulsions can also be sensitized for the infra-red, as far as 13,000 A.U. having been reached.

The sensitizing effect of erythrosine has been shown to be due to adsorption of light by the silver salt of the dye on the surface of the silver bromide. Each quantum absorbed frees a bromine ion, leaving the corresponding free silver atom in the lattice. The liberated bromine is taken up by the dye which is progressively bleached. In the presence of any other bromine acceptor, of which gelatin is one, the dye passes the bromine on, and is itself regenerated. Maximum photolytic formation of silver and maximum photographic sensitivity occur at the same concentration of dye, which is, however, considerably less than saturation. This suggests that part only of the silver bromide surface should be covered by dye. When entirely covered, sensitivity is only about one half of the maximum.[1]

Sensitizing dyes

Eosin and erythrosine were the first dyes commonly used for orthochromatic sensitizing. The formula of eosin, tetra-brom-fluorescein, is

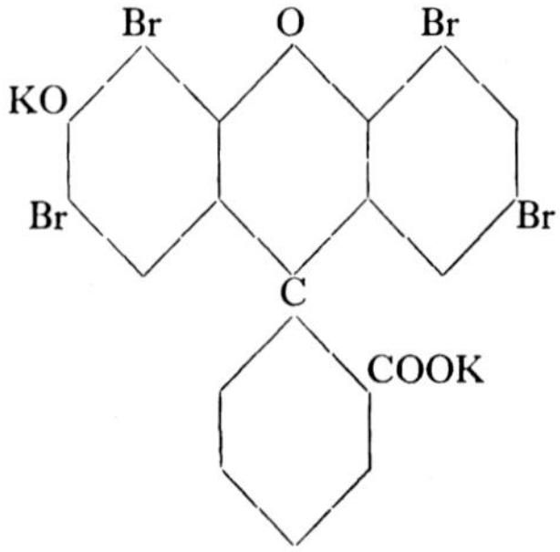

Br O Br
KO
Br Br
C
COOK

Erythrosine is the corresponding tetra-iodo-fluorescein.

[1] *J. Chem. Physics*, 426, **7**, 1939. S. E. Sheppard, R. H. Lambert and R. D. Walker.

The cyanine dyes are of special value as photographic sensitizers. Their chemistry looks alarming at first and their preparation and the elucidation of their structure is complicated, but now that these have been worked out their formulation is quite simple. The simplest member is formed by condensation of quinoline, I, and lepidine, II, which is 4-methyl quinoline, the two molecules being linked by the CH group from the original methyl group, III.

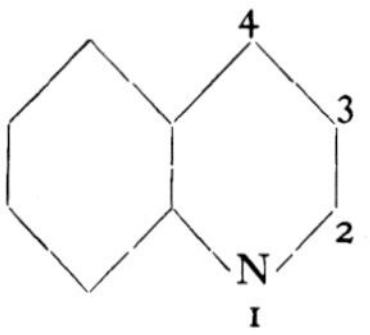

4 3 2 N 1

I

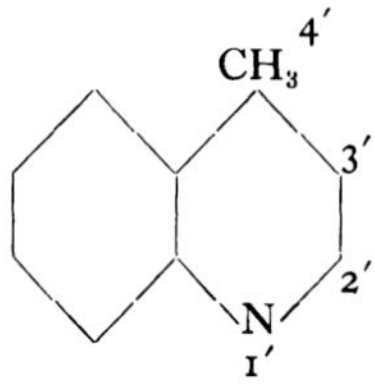

CH_3 4′ 3′ 2′ N 1′

II

R—N =CH— N R x

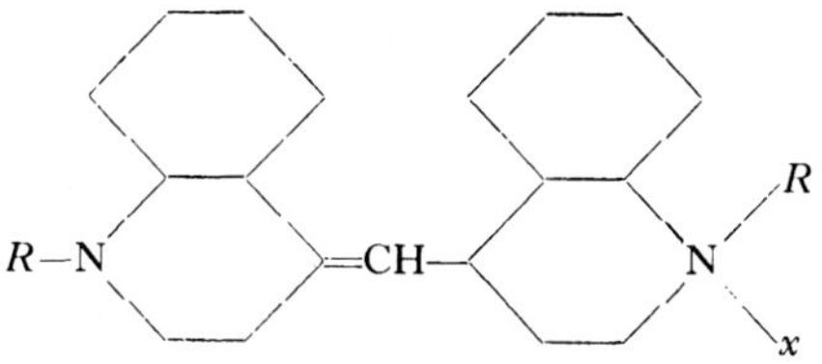

III. 4 : 4′ (true) cyanines

As can be seen, isomeric combinations are possible, and from 2-methyl quinoline—quinaldine—the pseudocyanines are formed, IV. In addition there exist the isocyanines, which are the 2 : 4′ compounds, position 1 and 1′ being, of course, the nitrogens.

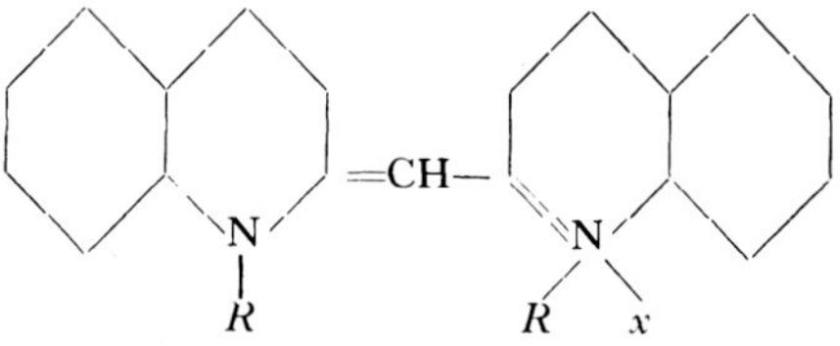

N R =CH— N R x

IV. pseudocyanine (2, 2′)

When the two ring systems are separated by longer carbon chains, we have the carbocyanines, which are named by the Greek numerals for the number of extra carbon atoms *in each half of the chain.* Thus the tri-carbocyanines have a 7-carbon chain, the tetra- have 9 and the penta- 11 (see formulae VII and VIII).

The carbocyanines also fall into three classes, similar to those of the parent cyanines:

2 : 2′	carbocyanines :	Pinacyanoles.
2 : 4′	,,	Dicyanines.
4 : 4′	,,	Kryptocyanines.

In addition, oxygen and sulphur can be introduced to form the oxacyanines:

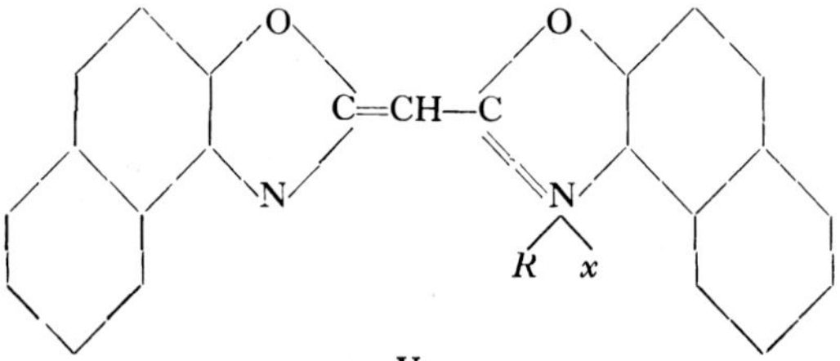

V

and the thiopseudocyanines:

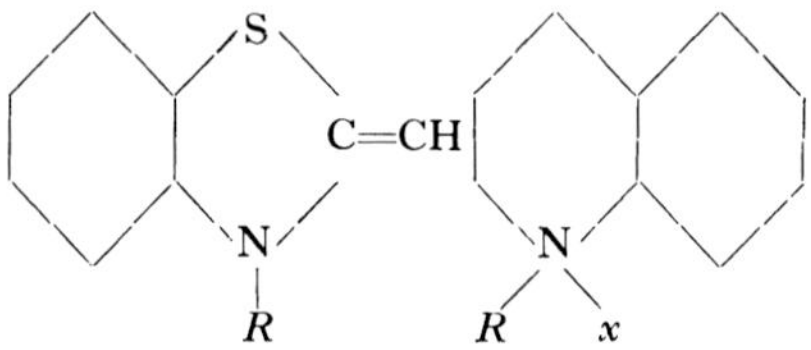

VI

It is of interest to note that increasing the chain length of the thiocarbocyanines extends their sensitivity farther and farther into the infra-red, 2 : 2′-diethyl tri-carbothiocyanine being a good example:

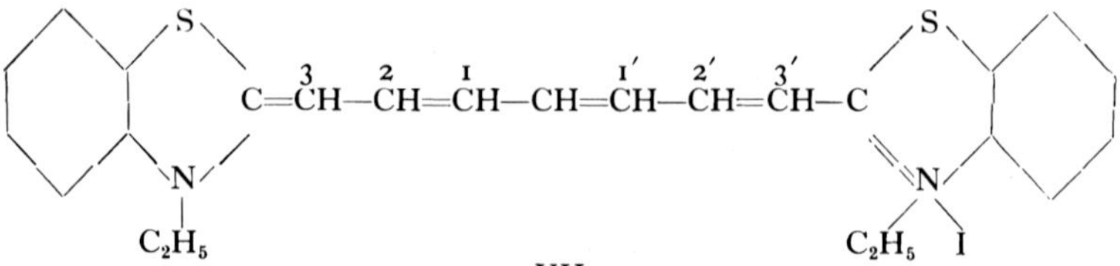

VII

This sensitizes up to about 13,000 A.U.

CYANINE

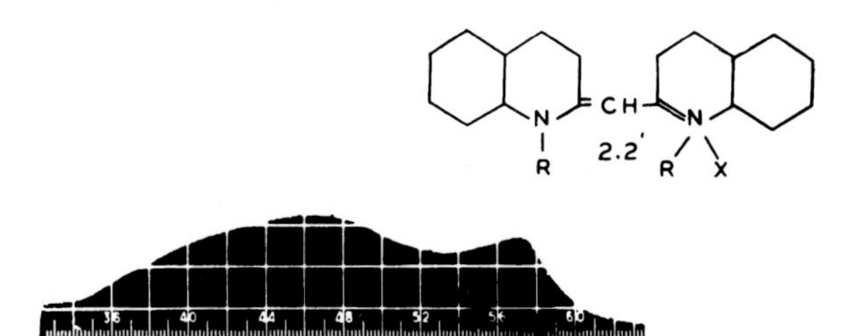

CARBOCYANINE

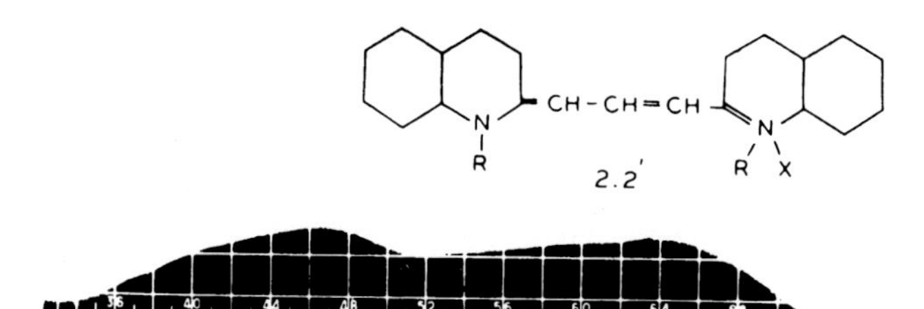

MESOCYANINE

CH–CH=CH–CH=CH–CH=CH

N

R

N

R X

2.2'

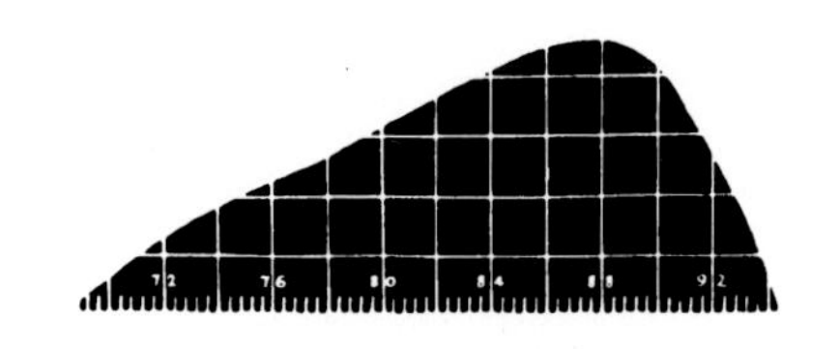

ultra-violet	violet	blue	green	yellow	red	infra red

Fig. 9. Spectrograms of cyanin dye sensitizers.

Finally we can write a general formula of the type:

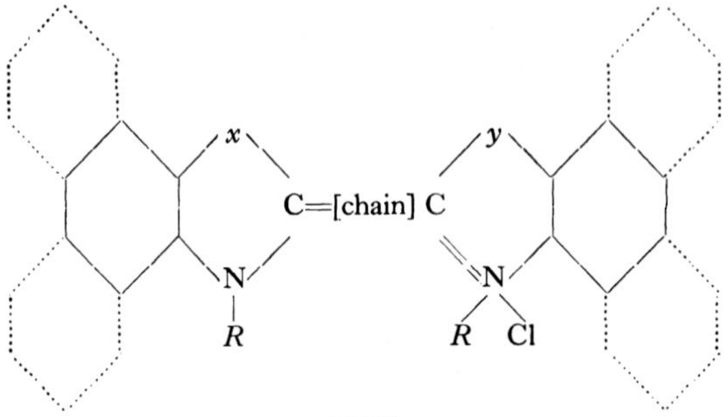

x y
C=[chain] C
N N
R R Cl

VIII

where x and y are oxygen, sulphur or selenium, or CH=CH; R is an alkyl group. The Cl can be replaced by other anions. Chains are:

=CH—	cyanin
—CH=CH—CH=	carbocyanin
=CH—CH=CH—CH=CH—	di-carbocyanin

and so on. In addition to the symmetrical cyanines, there is a still larger class of unsymmetrical ones.[1] Another variation is provided by isomerism in the carbocyanin chain. Alkyl groups may replace hydrogen. A special case of this is the neocyanine family, in which a third quinoline nucleus is attached to the centre carbon atom of the chain:

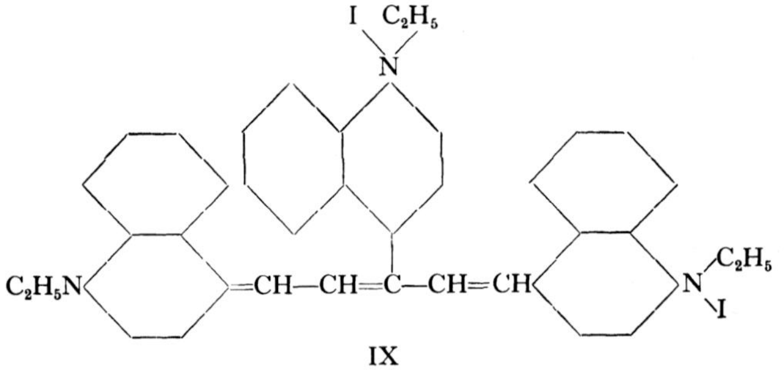

I C_2H_5
N
C_2H_5N =CH—CH=C—CH=CH N C_2H_5 I

IX

Fig. 9 shows the regions of the spectrum sensitized by three different cyanine dyes. I am obliged to Messrs Ilford for these photographs. They show well the shift towards the infra-red when

[1] See B. Beilenson, N. I. Fisher and F. M. Hamer, *Proc. Roy. Soc.* A, 1937, **163**, p. 138; W. H. Mills and R. C. Odams, *J. Chem. Soc.* 1924, p. 1913.

Fig. 10. Landscape photographed by infra-red light.

the chain length is increased. Mesocyanine is a product of Messrs Kodak.

The particular value of infra-red photography is described on p. 159, but it has one peculiarity which is often of great importance. Green foliage appears in the print snowy white. This is because chlorophyll does not absorb the infra-red in the range of wave-lengths usually employed. Fig. 10 was taken in the spring but, at first sight, looks like a snow scene. The blue sky is darkened but not the green leaves. They are lightened.

Correction filters

Filters are used either to correct, so as to get a photographic reproduction which approximates closely in monochrome to the visual impression, or for the exactly opposite purpose of increasing contrasts. Filters are almost invariably used just in front of the lens. This has the advantage that they are quite small. The most convenient is a sheet of gelatine containing a suitable dye. For permanence these can be cemented between glass plates. Coloured glass is also used, but the preparation of glasses which absorb, as required, is very much more limited than in the case of dyes. The absorption of light by the filter may be considerable when the glass is coloured by colloidally dispersed substances; e.g. cadmium sulphide for yellow glass. For scientific purposes it is sometimes more convenient to use as a filter a flat-sided glass cell containing a solution of the desired colour.[1] Plates containing yellow dye, to obviate the need for a filter, are also on the market. Such method of filtration is less efficient than an external filter.

Fig. 11 (*frontispiece*) shows photographs of a soap film taken by reflected light on the following materials: (*a*) is from a Dufaycolor transparency; (*b*) is on panchromatic material through a filter to give correct tone rendering of the colours; (*c*) is on a process plate; (*d*) is through a Wratten number 74 which transmits only a narrow band in the green; (*e*) is on a panchromatic plate through a red 10 times filter (Ilford tricolour red). It should be noted that the colours of the soap film are not pure spectral colours.[2] Comparison of the Dufay transparency with the original showed some de-

[1] See p. 47 for effects on optical system.

[2] For optical theory of the formation of the colours, see A. S. C. Lawrence, *Soap Films*.

gradation of the yellow band of the third order, which should be exceptionally brilliant. The most surprising weakness was the blues. They are bright enough in the transparency, but a light blue

Fig. 12. Penetration of haze using orange-red filter.

and not the brilliant deep indigo that they are in the soap film. Nevertheless, the gradual transition of colour in the wide band of the second order from carmine to violet, blue, blue-green is magnificently rendered. It may be noted that, the nearer the light is to monochromatic, the farther into the thicker part of the film can interference bands be detected.

Fig. 12 was taken through a micro-red filter (10 times exposure on panchromatic plate[1]) from the top of the Eiffel Tower. Visibility is much better than visual, the Sacré-Cœur church being seen only hazily and nothing beyond. The rendering of the foliage may be

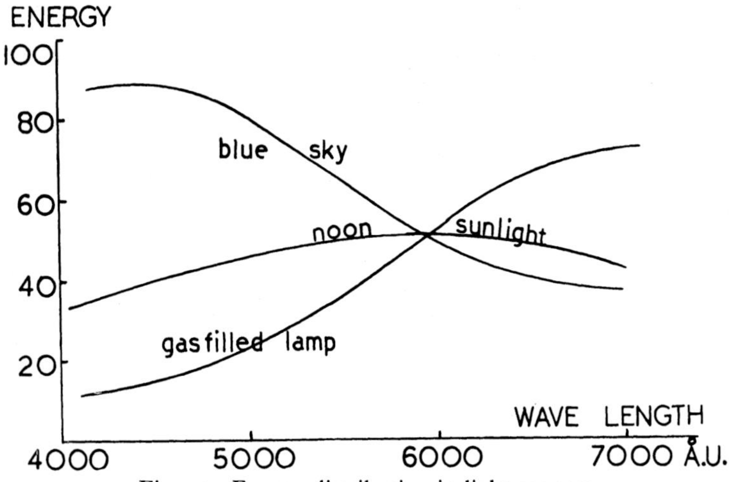

Fig. 13. Energy distribution in light sources.

compared with that in Fig. 10, taken by infra-red rays only. When using panchromatic plates the speed will vary according to the colour of the light in a quite different way from the variation in an ordinary plate. In particular the falling off of speed with yellow light—e.g., towards sunset or in artificial light—will be much less. Alternatively the speed in such coloured light will be much greater if the speed of ordinary plates is taken as standard.

Fig. 13 shows the energy distribution through the spectrum of a number of sources. The very much greater speed of a panchromatic emulsion compared with ordinary in $\frac{1}{2}$-watt light is obvious (see also p. 16).

Resolving power of emulsions

There is a definite limit to the resolution of any plate. Fortunately there is also a limit to the resolving power of the human eye. The

[1] It is customary to describe filters in common use by their colour and the increase of exposure required by a panchromatic emulsion. For light yellow filters, the factor for orthochromatic material is also given. For more exact requirements, the absorption spectrum must be used (see pp. 21, 101, 103).

limits of resolution by a plate are decided by several factors; the image of a point source is of finite size. The reasons for this (neglecting optical faults) are: (*a*) diffraction (see p. 51), (*b*) the size of the grains of silver in the developed emulsion, and (*c*) halation and light scattering in the emulsion.

The factors influencing grain are: (1) actual grain size of the silver halide, (2) grain growth in development, and (3) possible grain growth after development, i.e. during fixing and washing. As we have already seen, the most sensitive grains in an emulsion are the largest ones. Alternatively, if very fine grain is required, a

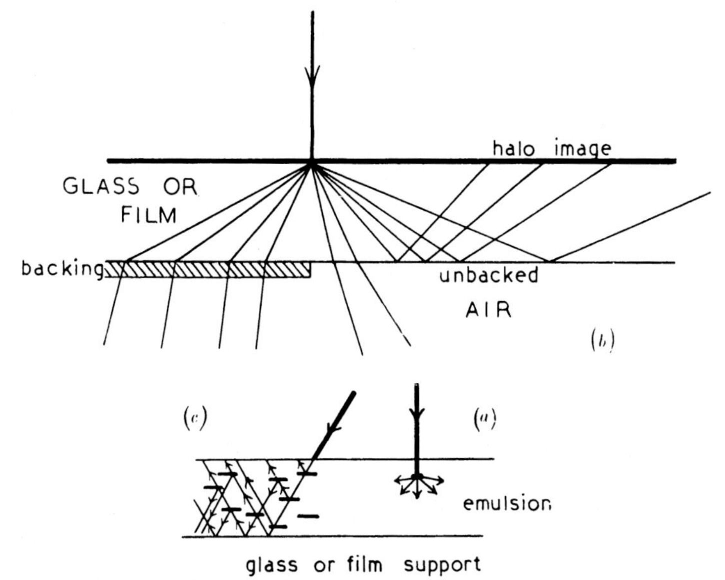

Fig. 14. Halation and light-scattering in emulsion.

slow emulsion should be used. It is now generally accepted that grain growth after development may be ignored and that grain growth during development is not serious in the sense of aggregation of grains to form very large ones. Strong developers, however, seem to give coarser grain by what Rabinowitsch calls 'explosive development'. Here the final grain may be considerably larger than the original silver-halide grains. There is still uncertainty on this problem, but it appears that keeping the alkali content low favours fine grain. See, for example, the Borax Metol

Hydroquinone developer given in Appendix I. The specific effects of developers are described in Chapter VI. An alternative method of obtaining fine grain is the so-called 'Physical development', which consists of depositing silver from a suitable solution upon the latent image after fixing. Extremely fine grain was claimed by some users of this method but it is rarely used.

Where high resolution is required special process plates should be used. Their speed is rather low so that, if a fast film must be used, the grain size can be checked only by the use of one of the so-called fine-grain developers. Fine-grain developers fall into two classes; the true fine-grain ones such as paraphenylene diamine, and the intermediate semi-fine-grain ones. The efficiency of these latter is still a matter for controversy. Their use is discussed more fully later (p. 124).

The resolution of an emulsion is also reduced by halation. Under this term two phenomena are often included. First, there is true scattering of light by the grains of the emulsion (Fig. 14 (*a*)).

Fig. 14 (*b*) shows true halation. Some of the incident light passes through the emulsion and is reflected internally from the back surface of the emulsion support. Obviously the radius of the halation disc will depend upon the thickness of the support, and halation is therefore much more troublesome in plates than in films. On a plate the image of a small point will appear to be surrounded by a ring which is separated from the spot by a ring unaffected by the halation. The critical angle for internal reflection is about 40°, so that nothing inside this limit is returned to the emulsion. Halation is prevented by coating the back of the emulsion support with a layer whose refractive index is as great as or greater than that of the glass, so that the light passes straight through without internal reflection. Mixtures of caramel and lamp-black are frequently used on plates. Alternatively, a layer of light-absorbing substance, such as a dye, may be interposed between emulsion and support.

Halation is complicated by scattering. This is shown diagrammatically in fig. 14 (*c*). It is not clear how far we are dealing with true scattering or with multiple internal reflection from the small plates of silver halide. Probably both phenomena occur according to the size of the particles, the smaller ones scattering and the larger ones reflecting. There is an important difference, since in true scattering, the shorter wave-lengths are scattered most.

The effects of grain size, halation and scattering upon sharpness of definition are discussed on p. 48.

Hypersensitizing

So far it has been shown that extra speed in an emulsion can be obtained by sensitizing with dyes, by ripening in the presence of potassium bromide which causes growth of larger crystals of silver halide, and by further ripening in the presence of substances which deposit grains of silver or silver sulphide. That these grains act as nuclei for development is clearly established. Obviously, therefore, any method of further increasing the number of nuclei should increase the speed of an emulsion. It may be noted here that intensification of a negative after development is not a method of increasing effective speed of an emulsion. It is usually assumed to be a method for making up for under-exposure, which amounts to the same thing, but it is nothing of the sort. An under-exposed negative, intensified, is still an under-exposed negative. All that intensification does is to increase the contrast which, in an under-exposed, fully developed negative, usually does more harm than good. The image, where there is any, is increased in density, but the blank parts on the negative (that is, those parts where the illumination is weak) remain blank. The only real value of intensification is where the negative is too flat to give a print which is sufficiently contrasty. This may be because the subject-matter lacked sufficient contrast, either because of inherent lack of contrast in the subject-matter or because of flat lighting such as occurs on a foggy day (see fig. 54). Or it may be due to under-development or to chemical fog in development. In all these cases, intensification is valuable. It should be noted that hypersensitizing will obviously increase the tendency to fog if development is carried far. Hypersensitization can be achieved by washing the plate in dilute ammonia, drying and using immediately. Much greater increase of speed can be obtained by exposing the emulsion to mercury vapour—*either before or after exposure*. It is sufficient to leave the film or plates in a wooden box containing a small globule of mercury for a period ranging from 4 to 48 hours according to the ease with which the vapour can reach the emulsion:

(*a*)

(*b*)

Fig. 15. Effect of mercury hypersensitizing.

that is, the longer period for films rolled up. Emulsions so hypersensitized lose their extra speed on standing for a few weeks. The gain in speed is equivalent to about 20 times and, for equal exposures, the increase of density about 1·3. The great advantage of this method is that hypersensitizing need not be used except on emulsion which is known to have received too short an exposure. Also, a film on which only one negative is under-exposed can be hypersensitized without spoiling the rest of the film. In fact, it has been suggested that it is rather improved, since the effect of hypersensitizing is to flatten out the characteristic curve so that with full exposure and hypersensitization, more correct tone rendering is given.

It is reasonable to connect the action of mercury with formation of new nuclei for development or with increasing the size of those nuclei which may be present already but which are too small to act as development centres. The non-permanence of the effect may be because the mercury is deposited as such and evaporates or because it reacts with silver halide to form silver and mercury halide which on standing dissociates again into mercury and halogen, the former of which evaporates while the latter reforms silver halide. Figs. 15 (*a*) and (*b*) show the effect of hypersensitizing.

These photographs were taken by Mr G. A. Jones, to whom I am also indebted for the following technical details. Both negatives were taken on the same spool of film (Agfa I.S.S.), one half being hypersensitized afterwards. Both halves were developed together in normal M.Q. developer for about 50 per cent longer time than normal to counteract flattening effect of hypersensitization. Print (*a*) is $\frac{1}{8}$ correct exposure (as indicated by Ombrux exposure meter) and print (*b*) is $\frac{1}{4}$. (*a*) is from the hypersensitized negative; (*b*) from the unhypersensitized. The density range in (*a*) was 0·61 to 1·50 and that in (*b*) was 0 to 0·89. The exposures were altered by stops and not by shutter speeds. The prints are the best that could be made from the two negatives. (*a*), with half the exposure of (*b*), is clearly the better picture as a result of hypersensitization and, indeed, would pass as a print from a normally exposed negative. whereas (*b*) can be seen by any one to be from an under-exposed one. The subject is somewhat dull because of the actual lighting conditions.

CHAPTER II

THE LENS AND THE IMAGE

Perspective and lens

The object of the lens is to form a sharp image of the object upon the light-sensitive material. When the object is in focus we have the relation $1/i+1/o=1/f$, where i is the distance between the image focal plane (the negative) and the lens, o is the distance of the object plane, and f the focal length of the lens.[1] Now since the object of photography is usually to obtain a picture resembling that seen by the eye we must have some limit to the angle of view included on the negative and, further, there will be a proper distance from which any positives must be viewed. If these conditions are not fulfilled we get incorrect perspective. It is taken conventionally that the focal length of the lens shall be approximately equal to the diagonal of the plate. This gives an angle of view of about 53°. That of the eye is of this order but varies with the conditions of vision. It follows now that the print of the same size as the negative must be viewed from a distance equal to its diagonal. This is impossible for the human eye when the diagonal is less than about 6 inches. We have here the explanation of why small pictures, which have to be viewed from a distance considerably greater than their diagonal, are so much improved by enlargement to a size which allows them to be viewed from the proper distance. This limitation obviously makes it desirable that the whole of the negative should be used for enlarging, otherwise one does not know the correct viewpoint and false perspective results. A very good example of this false perspective is provided by pictures taken with telephoto lenses of focal length long compared with the diagonal of the negative. For example, photographs of cricket matches shown in cinemas give an absurd perspective, since the angle of view for the audience is wrong. The distance of the audience from the screen, to give correct perspective, would be their normal distance multiplied by the ratio of focal length of telephoto lens used to that of the normal focal length which gives something approaching proper perspective. If the audience could

[1] See Appendix II for formal treatment.

move back to the distance at which perspective was correct they would lose all the advantages of the magnification of the image and would see it as they would have seen it with their own eyes from the edge of the cricket field or from where the picture was taken. It must be emphasized that photography cannot give in itself incorrect perspective. It can only give perspective which appears to be incorrect when the picture is viewed from the wrong distance. Unless the conditions mentioned—equality of focal length of taking lens with diagonal of plate, use of the whole negative in enlarging and inspection of the print from the distance equal to its diagonal—are fulfilled, false perspective may appear. In cases where there is no indication from the picture of the correct distance for viewing, as in the case where only part of the negative has been enlarged, our own experience sometimes helps to give the necessary correction.

In construction of pictures, whether for artistic or record purposes, perspective is of fundamental importance because it is the basic means of obtaining a three-dimensional impression from a two-dimensional picture. As already mentioned in the Introduction, the object of pictorial photography is almost invariably to make a record in two dimensions which gives the impression of the three-dimensional original.

The human eyes might well have been constructed for misleading the photographer. As we have seen (fig. 8 (*a*)), their response to the visible spectrum is quite different from that of any plate. From the point of view of reproduction of an object seen, it must always be remembered that the image seen is by binocular vision, whereas the camera has but one eye, which greatly increases the difficulty of showing three dimensions. It is very useful, when viewing a subject, to close one eye. It is often found then that subjects which appeared, with both eyes open, well arranged and separated now appear flattened out like a painted back-cloth.

Image formation and depth of focus

The human eye cannot distinguish objects smaller than a certain size and a small disk is indistinguishable from a point. Fig. 16 shows the geometry of image formation. The object plane, o_f, is focused on the image plane, i_f. Objects from nearer and more distant planes, o_1 and o_2, have their foci in the image planes, i_1 and i_2.

The images of the points o_1 and o_2, in the focal plane, are disks of finite size. It can be seen from the diagram that, other conditions being the same, the size of the disk in the focal plane will depend upon the diameter of the lens aperture. Now the equation

$$\frac{1}{f}=\frac{1}{o}+\frac{1}{i},$$

suggests that objects in one plane only are rendered sharply in the focal plane. But the image disks of objects in planes before and

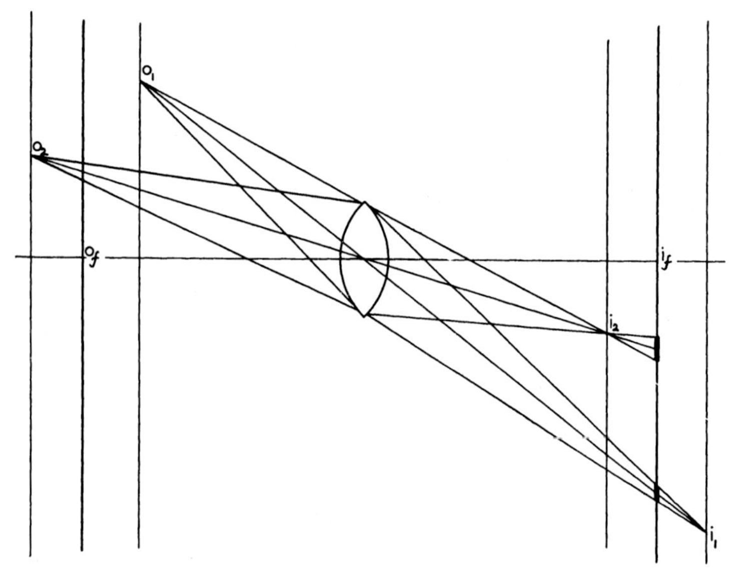

Fig. 16. Geometry of image formation.

behind the principal object plane will be indistinguishable from points in the image plane provided their diameter does not exceed a certain value. The size of this disk depends upon the distance from which it is observed. The solid angle subtended by the largest disk which is seen as a point is about 0·00029 radian. If pictures are always going to be viewed from the distance required for proper perspective, we may set a definite fixed upper limit to the disk. This is called the circle of confusion. The usual value is $\frac{1}{100}$ inch on a print of 10 inch diagonal. Before discussing the manner in which we can alter and control depth of focus—that is, total depth before and behind the object plane which appears in sharp focus—

the difficulties of constructing a lens and the sources of error must be discussed. It should be noted incidentally that this limitation of depth of focus is a very valuable tool to the photographer for giving the impression of the third dimension. By focusing sharply on the principal object of a picture and having everything before and behind it fuzzy an impression of depth is conveyed and the principal object is greatly strengthened.

The action of the lens: refraction and chromatic aberration

Light travels in straight lines provided that the optical density of the medium in which it is travelling remains constant. When it strikes a medium of different optical density it is refracted. Fig. 17

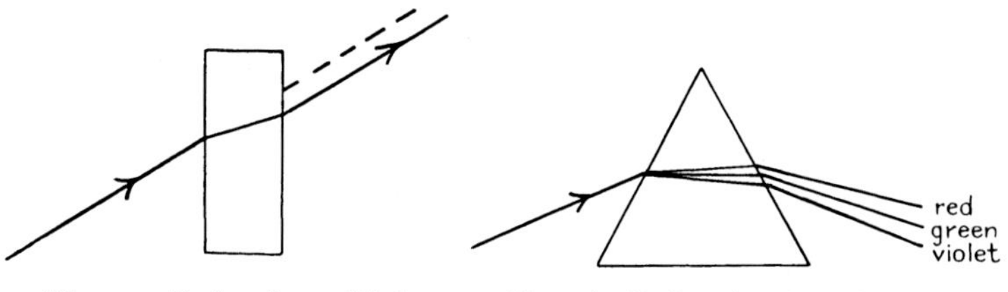

Fig. 17. Refraction of light by glass plate.

Fig. 18. Refraction by prism and dispersion of colours.

shows the refraction in a medium with parallel sides. Fig. 18 shows refraction by a prism. When white light is refracted by a prism it is split up into the spectrum because the refractive index is different

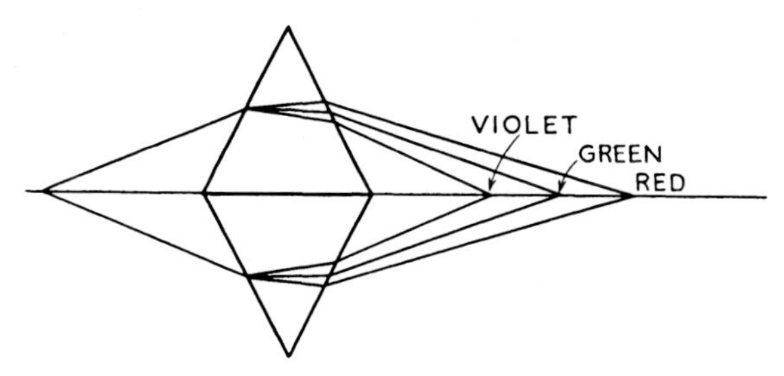

Fig. 19. Dispersion and origin of chromatic aberration.

for the different colours. If now, we consider two prisms placed base to base (fig. 19), we see that the beams diverging from a source are brought to a focus on the other side of the prisms; and that the focus for the violet end of the spectrum is closer to the prisms than is the red end. Now a lens may be regarded as made up of a very large number of sections of prisms in which the height of the section has become vanishingly small. We therefore see that the image of a point source of white light will not be a point but a series of points of different colour strung out at different distances along the optic axis. This is known as chromatic aberration. It is obvious from fig. 18 that another prism of the same shape and glass as the first might be placed upside down in the emergent beam and would gather together the different colours, which would then emerge

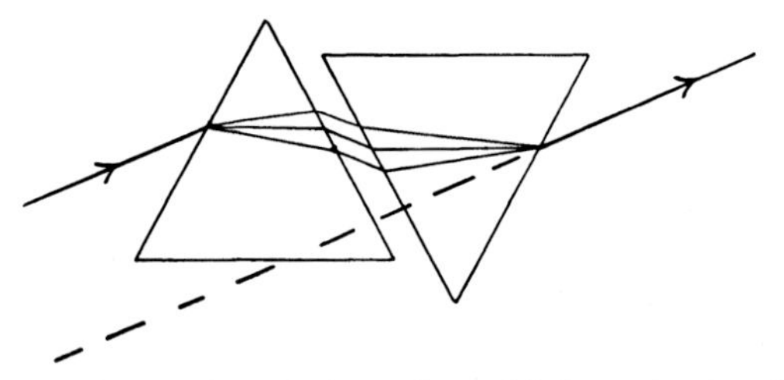

Fig. 20. Correction of dispersion with loss of refraction.

as a single white ray (fig. 20). It would, however, be parallel to the incident ray. The only way of evading this difficulty of getting refraction without chromatic aberration is to employ for the second prism a glass whose dispersion is different from that of the first. Dispersion of a material is the variation of refractive index with wave-length. In this way chromatic aberration can be reduced without losing refraction. It is usual for photographic lenses to be corrected for the D and G Fraunhofer lines (yellow and blue). The lens is then called achromatic. It should be noted that such lenses are not fully corrected for other colours, although the error due to this cause is small, compared with that of a single lens. The infra-red correction may be bad, but it so happens that in some lenses, e.g. Ross Xpress and the Cook Aviar, the visible and infra-red foci coincide.

Curvature of field

Curvature of field is easily understood if we consider the image of a plane at infinity focused by a sphere of glass. The geometry of image formation shows that the images lie on a spherical surface concentric with the sphere and at a distance from it equal to the focal length of the lens. Fig. 21 illustrates curvature of field. It is obvious that it would not be very serious in photographing open country, but with a flat object, such as a diagram, the edges will be out of focus when the centre is sharp and conversely. The considerations for depth of focus already outlined show that the effective limitation to this aberration is given by considering the

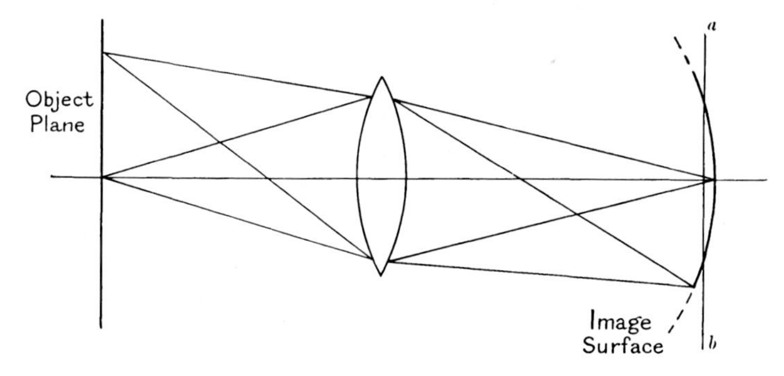

Fig. 21. Curvature of field.

plane *ab* in fig. 21. The axial rays will come to a focus beyond this plane and the marginal rays will be focused nearer to the lens. So long as the images of points on *ab* are not larger than allowed by the limiting circle of confusion, the whole image on the plane *ab* will be effectively sharp. It is clear, therefore, that curvature of the field imposes a limit to the useful angle of the lens.

This aberration also causes another peculiar effect, called *Distortion*, due to the position of the diaphragm. The image of a square appears on the plate as either a 'barrel' or 'pincushion'. It can be seen qualitatively from fig. 21 that the rays cut off by a diaphragm are very different according to whether the diaphragm is in front of or behind the lens. If the diaphragm is placed in the plane of symmetry of a compound lens, the aberration should disappear and, subject to certain limitations, does so.

Spherical aberration

Fig. 22 shows the origin of spherical aberration in convex and concave lenses. It can be seen that this aberration is caused by the marginal pencils, which come to a focus closer to, in the one case, and farther from, in the other, the focus of the axial rays. It is also obvious qualitatively that a combination of convex and concave lenses can be made which will correct this aberration. As with all other aberrations complete correction is not possible practically and the usual practice is to take two points, one on the axis and one

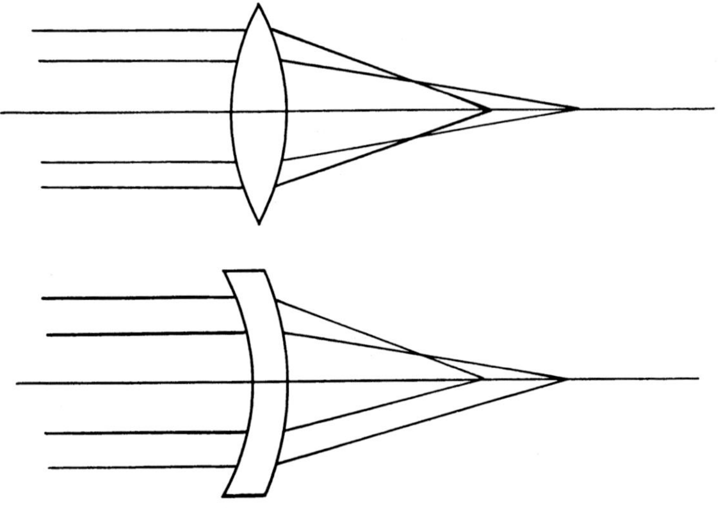

Fig. 22. Spherical aberration.

marginal point, and to correct for them. The shape of the curve connecting residual spherical aberration with distance from the axis will then depend upon the position chosen for the marginal point completely corrected. The position chosen for this depends upon the nature of the work for which the lens is required. It is also obvious that there will be secondary chromatic aberration due to differences of spherical aberrations of different colours.

Coma

Coma is an aberration of the marginal rays due to differences of refraction. It may thus be regarded as spherical aberration of the extra-axial rays. Fig. 23 shows the origin of this aberration. The result is that the oblique rays form an image of a circular object with a conical extension. The angle of this cone is 60°. This is one of the most serious aberrations, since the loss of definition of an object is not symmetrical and, if measurements are to be made from a plate, it is impossible to see what is the true centre of the object.

The extent of coma increases as the square of the aperture and directly as the distance of the image from the optic axis. Abbe and others have shown that freedom from coma requires that

$$\frac{\sin u}{\sin u_1} = \frac{N_1}{N} M,$$

where u and u_1 are the angles subtended by object and image and N_1 and N are the refractive indices of object and image spaces. M is the magnification.

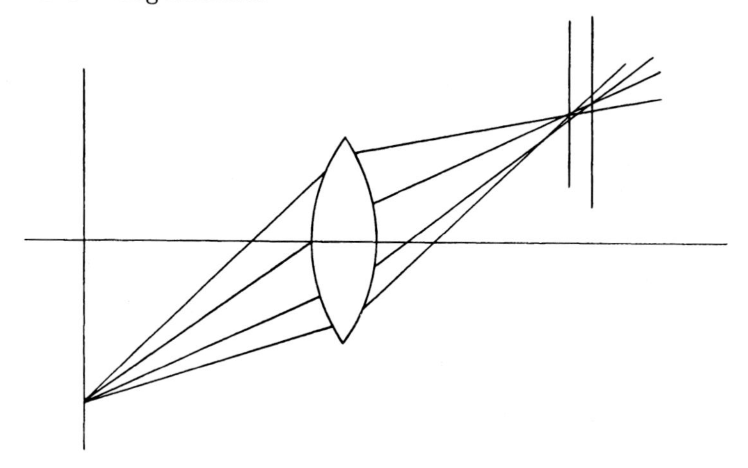

Fig. 23. Coma.

But correct image formation requires that

$$\frac{\tan u}{\tan u_1} = \frac{f_1 + X_1 f}{f + X f}.$$

Under any given set of conditions the right-hand sides of these two equations are constant for any given object, O. It is necessary, therefore, that the ratio of the sines and tangents of u and u_1 shall be equal. If we want our lens system to be perfect for all object distances it follows that we are limited to angles such that the difference between sin and tan is negligible. The difference is very small up to about 4°. This angle corresponds with a lens working at about $f/7$.

Astigmatism

This is one of the most serious of the aberrations of lenses and most difficult to correct. It occurs in the light rays passing through the lens obliquely, but differs from spherical aberration in that it does not affect the axial rays whereas the latter does. Astigmatism is due to asymmetry of refraction in different planes in the lens.

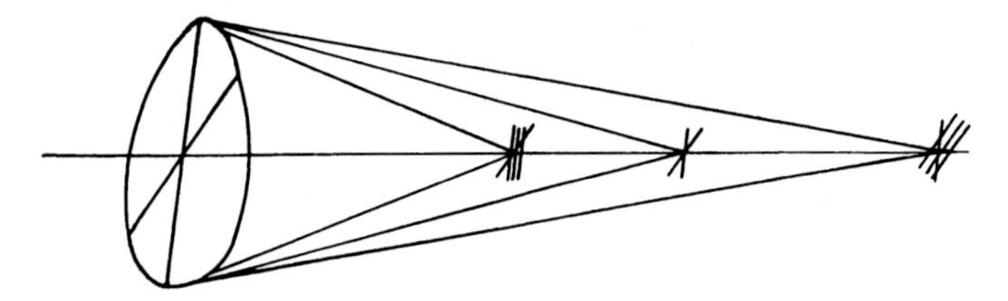

Fig. 24. Astigmatism.

If we consider the image of two lines crossing at right angles, we find that there is a difference of the image distance from the lens for the two lines. In other words, it is impossible to get both in sharp focus at the same time. This is shown simply in fig. 24.

Flare

Flare spots occur sometimes as circular patches of diffuse light in the centre of the picture. These are due to internal reflections inside the lens. Since it originates from the diaphragm, being an image of the aperture close to the focal plane, the effect is worst for small stops. It may be eliminated by moving the diaphragm but loss of definition will result.

Performance of a lens

The rather terrifying list of aberrations need not be taken too seriously since there are two ways in which their effects may be reduced so far that loss of sharpness is within the circle of confusion permissible. The aberrations described are for single lenses. Compound lenses, built up of several components of glasses of different characteristics, are supplied by numerous manufacturers. These, although not theoretically perfect, are sufficiently corrected for all but exceptional jobs. The performance of a lens may be tested by focusing the camera on an object, such as small print or ruled lines, at a distance, removing the back of the camera and examining the image in the focal plane with a microscope. It will be found usually that the resolving power of the lens is considerably greater than that of the emulsion.

Stops

It will have been noticed that, in most cases, the rays of light traversing the peripheral parts of the lens are those responsible for the worst aberration. Restricting the path of light to the axial and near-axial rays, therefore, improves the performance of the lens very considerably. Originally this was effected by metal plates with holes of suitable size, named, for obvious reason, 'stops'. Now, iris diaphragms in which the size of the stop can be continuously varied are almost invariably used.

Chromatic aberration is proportional to the size of the stop; spherical aberration to the cube and astigmatism also directly to its size. Curvature of field increases approximately with its square.

F numbers

If we consider a camera focused on to a point source of light at a distance, O, then the amount of light reaching the lens is, by the inverse square law, A/O^2, where A is the area of the lens aperture. The total amount of light entering the lens from any plane is proportional to the area of that plane included in the field of view of the lens; that is, as O^2. The amount of light passing through the lens is therefore proportional to the area of its aperture or to the square of its diameter, D. The image size (linear), however, is proportional to the focal length of the lens and its area to F^2. The intensity of light on the image plane is therefore proportional to D^2/F^2. If now we define D as a fraction of the focal length F/n, then the light intensity is F^2/n^2F^2, or as $1/n^2$. A conventional series of numbers can be chosen to increase by $\sqrt{2}$, so that each rise in the series requires a doubling of the exposure. Such a series means that for any value of n, the light intensity and the exposure will be the same whatever the focal length of the lens or distance of the object. This breaks down only for objects very close to the camera, since the size of the image, using any given lens, is equal to the size of the object multiplied by $\frac{F}{O-F}$. Except for very small object distances, O is much greater than F, so that the latter may be ignored and the relation given above is accurate. For very close objects, however, F becomes appreciable with respect to O and, as the image becomes larger, the light intensity is reduced as the square of $\frac{F}{O-F}$. This is the reason why close-up subjects need extra exposure.

The conventional series of stops or F numbers, as they are usually named, is as follows:

1, 1·4, 2, 2·8, 4, 5·6, 8, 11, 16, 22·6, 32.

Common maximum apertures for lenses are $F/3{\cdot}5$ and $F/4{\cdot}5$, which obviously fall between the strict series. Usually, however, there is nothing on the lens markings to indicate that these maximum apertures are not twice as fast as the next higher F number, the conventional series being used for all but the maximum.

Other markings are sometimes found on very old lenses. If there is any doubt, a re-calibration can be made knowing the focal length of the lens and the diameter of the stops. It should be noted, however, that the true diameter of the aperture is not the *effective aperture*, since the light, after passing through the front combination of the lens, becomes convergent, so that more light passes through a stop of given diameter than if the rays were parallel. In single-lens cameras, the real and effective diameters are the same if the diaphragm is in front of the lens.

Depth of focus

Fig. 16 shows the principle of depth of focus. The importance of this factor in photography cannot be over-emphasized. It also embodies the basic difference between visual and photographic sight. The eye focuses on a plane; before and beyond is out of focus, but the eye alters its focus gradually as required to take in any given depth of focus by an integrating process. The camera, on the other hand, can be adjusted to give much greater depth of focus if required. But the camera, having focused on a given object plane, takes a picture in which a certain depth is in sharp focus and the rest out of focus. Nothing can alter this distribution. With the eye, on the other hand, the chosen object plane can be altered as often as desired, and this alteration is usually subconscious.

There are several ways in which the depth of focus, d, can be calculated, and the reader is referred to the textbooks on Optics (p. vii). The equation

$$d=\frac{2n\delta O\,(O-F)}{F^2},$$

gives total depth of focus in terms of stop number, n, object distance, O, maximum permissible circle of confusion, δ, and focal length of lens, F. Depth of focus is therefore directly proportional to the stop number (that is, inversely proportional to the actual aperture), inversely proportional to the square of the focal length

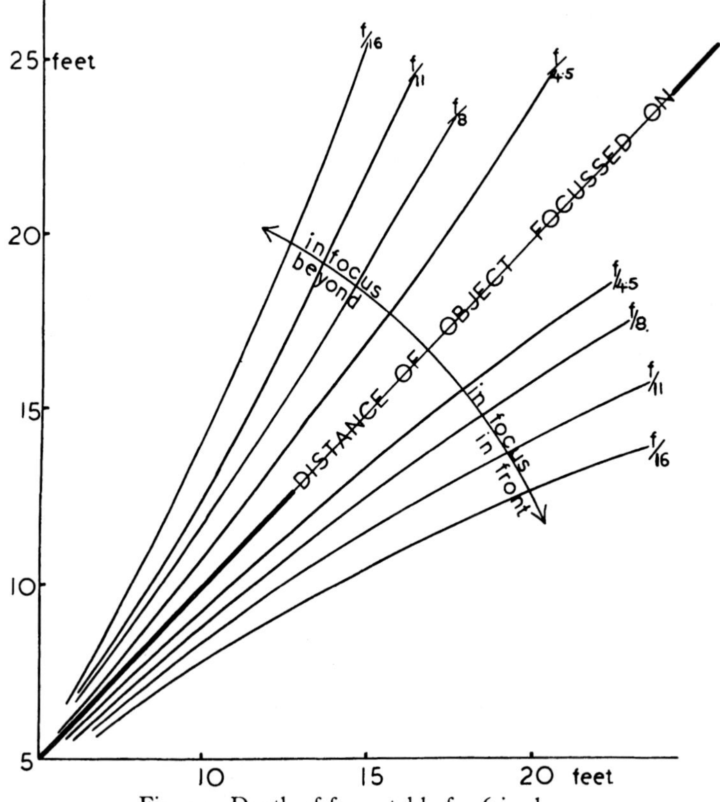

Fig. 25. Depth of focus table for 6 in. lens.

of the lens (for a stated value of the circle of confusion) and, for distant objects, proportional to the square of the object distance. It shows also that, as O decreases, the depth of focus becomes less than O^2. In other words depth of focus is least for the nearest objects. Since depth of focus is proportional to F^2, it is also proportional to the square of the diagonal of the negative for a normal angle lens, so that the smaller the negative size, the greater the depth of focus, in inverse proportion to the square of its size. What we really want to know, however, is the circle of confusion on a print of given size, irrespective of the size of the negative from which it was printed. The amount of magnification required by a small negative is obviously greater than that for a larger negative, in inverse proportion to the ratio of their sizes; that is, of the focal lengths of the lenses used. Instead, therefore, of taking a fixed value for δ we may substitute $F/1000$; the equation then becomes

$$d = \frac{2n\,O\,(O-F)}{1000F}.$$

Fig. 25 shows the depths of focus for a lens of focal length of 6 inches (for covering a quarter plate) at a number of stops. The values are calculated for a circle of confusion of $F/1000$.

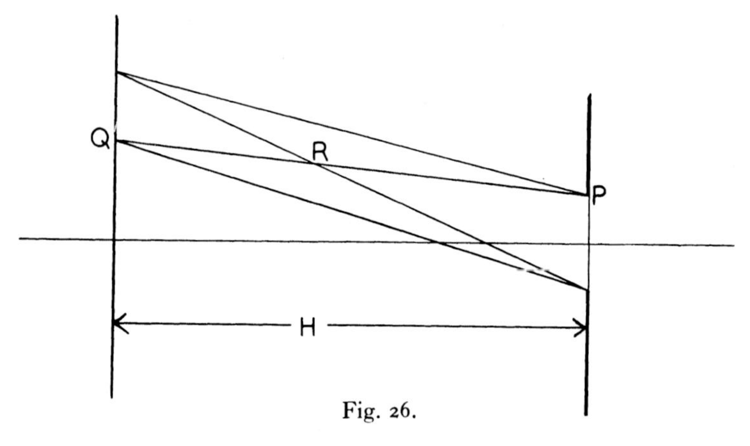

Fig. 26.

For the practical problem of determining depth of focus, it is often more convenient to adopt the following procedure. In landscape work the depth most frequently required is that between some object in the foreground and the horizon, i.e. infinity, hence the definition is applied to *hyperfocal distance*—the distance in focus beyond which everything is also in sharp focus. Nearer objects are also in sharp focus to a distance $H/2$. This distance, H, can be calculated independently of the expression for depth of focus, as follows: in fig. 26 P is the entrance pupil. Parallel rays of light (coming from a great distance) will cut planes perpendicular to the axis, and the image at Q will be a circle of the same diameter as the entrance pupil when the beam has its apex at R, half-way between Q and the entrance pupil. The condition now required is that, with the lens focused on Q, all objects from R to infinity shall be sharp within the limits of resolution of the eye.

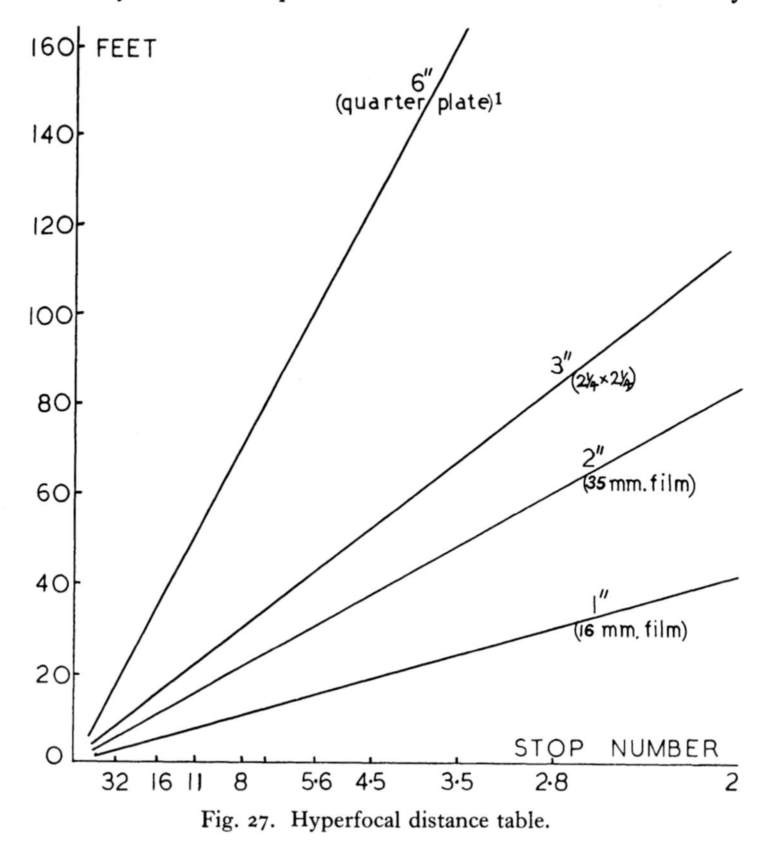

Fig. 27. Hyperfocal distance table.

That is, that c/m shall not exceed the diameter of the circle of confusion where c is the diameter of any of the images on Q. Since the aperture is equal to F/N, this equals

$$\frac{F}{mN} = \delta,$$

where m is the reduction factor.

[1] The English names for certain sizes of plates, etc. are (in inches): lantern slide, $3\frac{1}{4} \times 3\frac{1}{4}$; quarter plate, $3\frac{1}{4} \times 4\frac{1}{4}$; half plate, $4\frac{3}{4} \times 6\frac{1}{2}$; whole plate, $6\frac{1}{2} \times 8\frac{1}{2}$.

H is therefore $$\left(\frac{F}{Ne}+1\right)F.$$

Now $$D_1=\frac{HD}{H-D} \quad \text{and} \quad D_2=\frac{HD}{H+D},$$

where D_1 is depth of focus behind object and D_2 in front.

The depths of focus before and behind the plane focused on being $$\frac{D^2}{H+D} \quad \text{and} \quad \frac{D^2}{H-D}.$$

Total depth of focus is therefore
$$\frac{2HD^2}{H^2-D^2},$$

where H is hyperfocal distance and D is distance focused on.

Fig. 27 shows the hyperfocal distances for a number of the commoner sizes of lenses fitted to cameras to give the conventional angle of view.

Depth of focus tables are supplied by some manufacturers with their camera. Alternatively they may be calculated from tables such as that in the *British Journal Photographic Almanac*. It is very important that every owner of a camera should know the depth of focus at any stop and object distance. Otherwise he will complain of the results obtained at small object distances and blame the lens for lack of definition, which is, in fact, due to fundamental optical considerations alone.

Effect of filter upon focusing

It is commonly believed that filters may have serious effects upon the optical system. Provided, however, that the filter is free from gross optical defect, it has little effect upon the focusing or upon the sharpness of the image. The simplest way of testing the filter is to view on it the image of a light source, such as an electric lamp, and to compare the two reflected images from the front and back surfaces. Any defects become apparent as the filter is moved about. Fig. 28 shows the cause of the displacement of focus by a filter due to its finite thickness. The actual displacement of the focal plane is easily seen by considering the depth of air equal to a filter thickness of θ (refractive index μ), which is θ/μ. The difference is $\theta(\mu-1)/\mu$. For glass filters this is about $\theta/3$—a negligible amount. The displacement is then $\dfrac{\theta}{3m^2+\dfrac{\theta m}{f}}$, where $1/m$ is scale of reproduction. $\theta m/f$ may be ignored compared with m^2, so that we get for the displacement $\theta/3m^2$, which is clearly of no significance except where the object is being reproduced on a

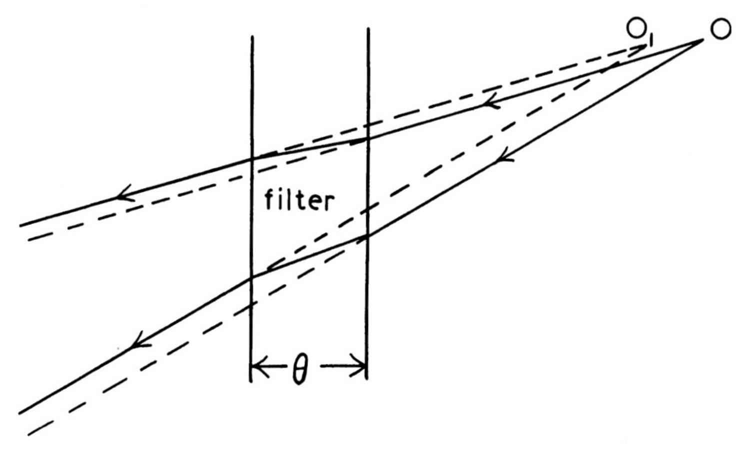

Fig. 28. Displacement of focus by filter.

magnified scale. The only case in which the error may be serious is when θ is large, as when a cell containing solution is used as a filter. Serious alteration of the focus may occur as a result of using a filter in infra-red work. Here the focus may be displaced considerably but the effect is an extreme chromatic aberration rather than an optical effect due to the filter.

Absorption of light by the lens

It is usually stated that all lenses of the same *f* number require the same exposure, other conditions being the same. This, however, is not strictly true owing to differences of amount of light transmitted through the lens. For this reason very cheap cameras fitted with a single meniscus lens are actually considerably faster than more fully corrected and expensive lenses used at the same nominal aperture. Neblette (p. 85) gives the following table for the loss of light in a number of lenses and the effective *f* value for the theoretical ones.

Loss of Light in Photographic Objectives

Objective	Marked *F*/value	Trans-mission	Effective *F*/value
Ernostar	*F*/2	46·6 %	*F*/3·1
Rüo	*F*/2	58·1	*F*/3·26
Kinoplasmat	*F*/1·5	38·7	*F*/2·82
Planar	*F*/4·5	52·2	*F*/5·65
Tessar	*F*/3·5	60·0	*F*/4·64
Tessar	*F*/4·5	63·5	*F*/5·94
Dogmar	*F*/4·5	52·7	*F*/6·39
Aviar	*F*/4·5	49·7	*F*/6·65
Collinear	*F*/6·3	53·4	*F*/8·67

Marginal illumination

When a flat image is formed of an object, it is obvious that the illumination at a point near the margin will be less intense than at the centre, since the margin is farther away from the lens. The light reaching a marginal point is not normal to the light-sensitive material. The combined result of these two effects is that the illumination falls off at a rate between $\cos^4 \alpha$ and $\cos^3 \alpha$, where α is the angle between marginal ray and optic axis. Fig. 29 shows the falling off for $\cos^4 \alpha$. It is obvious that a field of 36° ($\alpha = 18°$) involves falling off of only 20 per cent, a difference not perceptible as density difference on the negative except by accurate measurement.

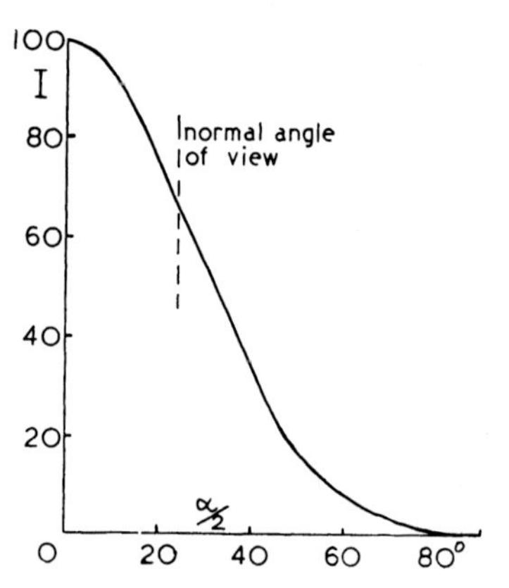

Fig. 29. Falling-off of illumination at edge of negative.

Sharpness and definition

When resolution and definition are of importance, slow emulsions should be used as far as possible since, in general, they have finer grain. If the lens is used properly according to the limitations described already, the limit to resolving power is the negative grain.

The edge of the image of a sharp edge will not be perfectly sharp on the negative on account of spreading. Aberrations of the lens may cause loss of definition but, in general, the residual aberrations of a well-corrected lens may be neglected and the loss of sharpness is due to the emulsion alone. The sharpness, S, may be considered in the light of the density gradient $\frac{d\Delta}{dx}$, where Δ is density and x is the distance from the true image edge. If we multiply by

$$\frac{d \log E}{d \log E},$$

we get

$$S = \frac{d\Delta}{d \log E} \times \frac{d \log E}{dx} \quad \text{or} \quad \frac{\gamma}{k},$$

where k is the diffusion factor. With any given emulsion for which k is constant, it is clear that the diffuseness follows the form of the characteristic curve of the emulsion. Fig. 30 (*a*) shows the amount of spread for three different densities of the true image. For maximum sharpness, it is clear that an emulsion of the 'process' type with a very steep gamma should be used. Fig. 30 (*b*) shows in curves 1 and 2 the spread for a given exposure with long and short development times respectively. Curve 3 is for a longer exposure but with development time cut down to give the same density in the true image as curve 2. The spread is less, which suggests that the common practice of giving long exposure and short development time to reduce grain in small negatives may be accompanied by more loss of definition than with less exposure and longer development. It is general experience that enlargements from miniature negatives show some loss of sharpness well before grain is apparent.

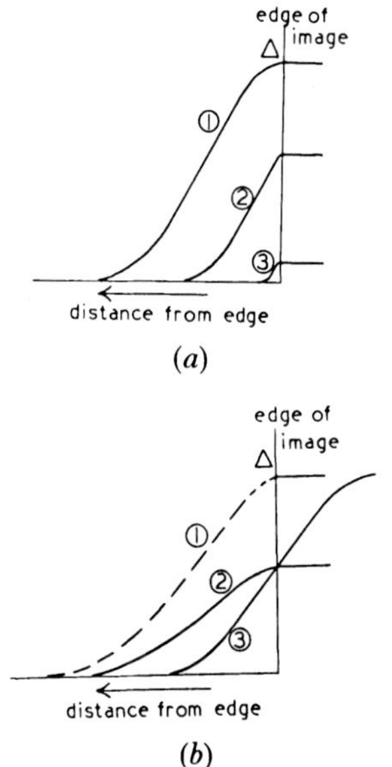

Fig. 30. Sharpness of edge of image.

Photography of line diagrams should be done on roll film or cut film, not on plates, since halation is then an additional cause of unsharp definition. It is also forgotten frequently that most diagrams are reduced considerably for printing. The width of the lines is therefore reduced to a half or a third. If there is any suspicion that chromatic aberrations are present, a blue filter should be used.

It should be noted that for maximum resolution of very small details, light of the shortest possible wave-length should be used (see p. 51) to reduce diffraction. In outdoor work diffraction is of less importance than scattering by small particles of dust and water in the air.

Aerial light scattering

This scattering is also a function of the wave-length but operates in the opposite direction—inversely as the fourth power of the wave-length. The Rayleigh equation for the intensity of scattered light, I, is

$$I=\frac{9\pi^2 n v^2 A^2}{\lambda^4 x^2}\left(\frac{\mu_2{}^2-\mu_1{}^2}{\mu_2{}^2+2\mu_1}\right)\sin^2\alpha,$$

where I is the intensity of scattered light, n is the number of particles per unit volume, μ_1, μ_2 are the refractive indices of medium and particles, v^2 the mean square volume of the particles, A the amplitude of the incident light and λ its wave-length, α is the angle between directions of vibration of incident and scattered polarized rays, x is a distance from the particle at which the calculation is made. This equation provides the explanation of the spectacular results obtained by infra-red photography. It also indicates the care which must be used in the neighbourhood of the sea, where there is much light-scattering matter in the air and an excess of light of the shorter wave-lengths, including ultra-violet.

Diffraction and resolving power

The image of a point source is a disk of finite size. Its size may be calculated in the following manner (fig. 31). A point source at infinity is focused on to the point F. At a distance f from the point F is a diaphragm whose aperture is of radius R. Now consider the illumination at the point P in the focal plane. For the optical paths we have

$$AP^2=f^2+(R+r)^2,$$

$$BP^2=f^2+(R-r)^2,$$

so that $AP^2-BP^2=4Rr$, and the path difference $AP-BP$ is

$$\frac{4Rr}{AP+BP}.$$

But AP and BP can be taken as $2f$ since r is small compared with f. The path difference is therefore

$$\frac{2Rr}{f}.$$

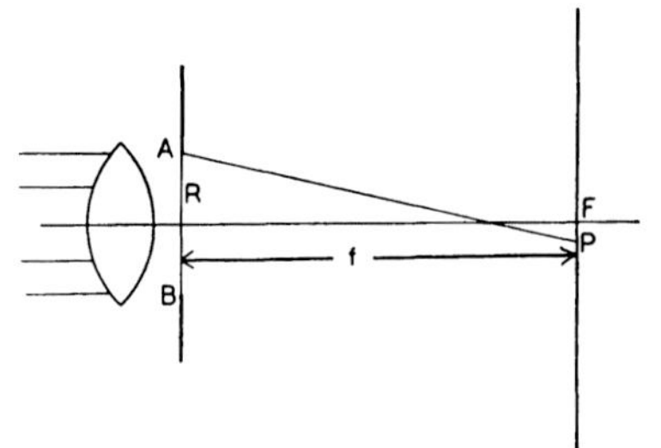

Fig. 31. Diffraction and sharpness.

P will be bright or dark according to the equation

$$\frac{2\,Rr}{f} = n\lambda \quad \text{or} \quad (n+\tfrac{1}{2})\,\lambda.$$

Since numerical aperture is defined as f/N, where N is the stop number, we have the relation

$$\frac{r}{N} = n\lambda,$$

or, for the first diffraction halo,

$$r = N\lambda.$$

For photographic purposes $2r$ must not be greater than the permissible disk of confusion, δ, so that $d = 2N\lambda$, or N must be smaller than $\delta/2\lambda$.

This treatment has so far considered monochromatic light. In actual practice using white light, by the time that the red diffraction image has reached its first maximum the shorter wave-lengths are taking its place around the image and coloured haloes are formed.

In microphotography, a careful watch must be made for diffraction. Diffraction is not an aberration of the lens. It results from the wave nature of light itself. The best lenses are therefore as liable to show the effect as poor ones. Coloured images are frequently seen on the ground-glass screen (in the apparatus shown in fig. 72) at high magnifications and then a filter should always be used. Blue or violet are the best, since the size of the diffraction image is smallest for the shortest wave-lengths. **Finally, to reduce diffraction, open up the lens aperture—the exact opposite of the procedure for increasing sharpness when this involves depth of focus.** Lord Rayleigh has pointed out that diffraction can be reduced by use of a solid circular stop to cut off the axial rays, since it is these that are mainly responsible for the diffraction image (fig. 31). In practice, however, this is rarely feasible, since it means using the extra-axial parts of the lens which are the parts producing the ordinary aberrations.

It will be seen from another form of the Rayleigh equation—

$$I_s = I_i \left(\frac{\mu_1^{\,2}}{\mu_2} - 1 \right)^2 (1 + \cos^2 \beta) \frac{nv^2}{a^2 \lambda^4},$$

where I_s is intensity of scattered light,

I_i is intensity of incident light,

μ_1 refractive index of scattering substance,

μ refractive index of dispersion medium,

β = angle between incident light and angle of view,

n = number of particles per unit volume,

v = volume of particles,

λ = wave-length of scattered light,

a = distance between particle and observer—

that no lower limit is set to the size of the particles scattering light, and molecules must be considered. The light from the sky is not only blue but also polarized. Both of these phenomena are explained by the equation. The amount of gas required for the effect is very large, since the molecules are so small and the scattering is proportional to the square of their diameter. That the molecules are responsible for the effects in the sky is shown by the fact that the polarization still persists above the level of dust particles.

Since the intensity of scattered light depends upon the square of the cosine of the angle between line of sight and direction of incident light, scattering is at a maximum when this angle is 90°. This obviously happens at midday and explains the much better definition of distant objects obtained in early morning or late afternoon.

Telephoto lenses

It is often desirable to increase the size of the image above that given by a lens whose focal length is equal to the diagonal of the negative. Incidentally, it may be noted that one of the commonest faults of the amateur photographer is to take things too small. The size of the image as already shown is given by the equation

$$\frac{S_i}{S_o} = \frac{i}{O},$$

where S_i is image size, S_o object size, i extension (distance from lens to image), O object distance. In other words, S_i increases with extension of lens.

It is not convenient, however, to increase the extension of the lens to values sufficiently large to give much increase in the size of the image. General-purpose cameras are often fitted for this purpose with a double extension which does not involve undue bulkiness (see fig. 35, p. 58). For the reflex camera or the box

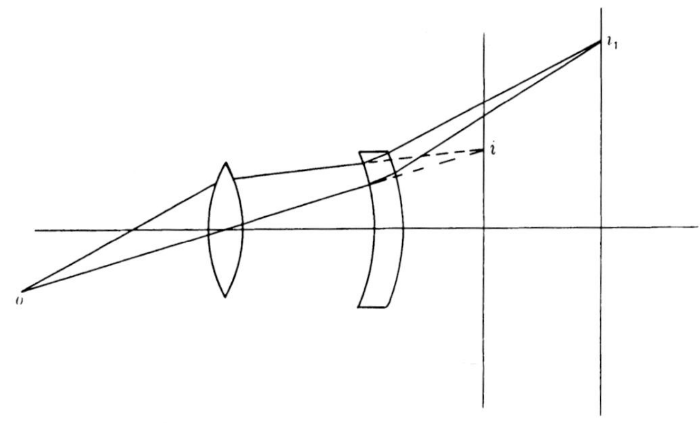

Fig. 32. Principle of telephoto lens.

camera whose extensions are limited the necessary increase of extension can be obtained conveniently by attaching the lens to a tube of suitable length which is screwed into the front panel. Care must be taken that this tube is of sufficient diameter to prevent cutting off the marginal rays. The telephoto lens acts on a different principle—the name telephoto is sometimes now misapplied to normal lenses mounted in extension tubes—namely, addition of an extra concave component at the back of the objective which alters the angle of the light after passing through the lens. Fig. 32 shows the principle. The rule for combination of lenses is given by the equation

$$s=f_1-f_2+\frac{f_2}{M},$$

where s is separation, f_1 and f_2 the focal lengths and M the magnification due to second lens (concave), and the resultant focal length of the combination F is given by

$$F=\frac{f_1 f_2}{f_1+f_2-s}=\frac{f_1 f_2}{\delta},$$

where δ is the optical separation of foci f_1 and f_2. The separation is strictly to be measured from the node of emission from the front lens to the node of admission to the back lens. It is obvious that the scale of the image varies with the separation between the components, other conditions being the same. This principle is made use of in the variable focus telephoto lens. With variable focus, the field included on the plate also varies with scale of magnification. For accuracy the fixed separation type is to be preferred. In calculating exposure for a telephoto lens the normal exposure required for the aperture of the lens and other conditions must be multiplied by the square of the magnification factor, i.e. the ratio of focal length of telephoto lens to normal focal length (equal to the diagonal of the plate) or by the ratio $(i_i/i)^2$.

Wide-angle lenses

It frequently happens in photography of architectural subjects that the whole subject cannot be obtained on the plate at the maximum distance allowable by surrounding buildings. In interiors a similar limitation is common. For this purpose a wide-angle lens must be used. The angle may be as much as 100°. For adequate covering power wide-angle lenses need to be stopped down considerably. Wide-angle lenses require small camera extension (p. 56), as they have short focal length. This prevents their being used with many cameras (p. 60).

Soft-focus lenses

Soft-focus lenses are often useful for artistic effects, but the difference from out-of-focus fuzziness is that a true soft-focus effect is a uniform slight diffusion over the image, while the definition in an improperly focused negative varies from sharp in the plane in sharp focus to complete fuzziness in planes before and behind it. The usual method employed, especially in portrait lenses, is an adjustment of the separation of the components of the lens which allows either an image of maximum sharpness or, after alteration of the separation, a slightly blurred effect due to spherical aberration. A soft-focus attachment can be obtained consisting of a sheet of glass, one surface of which is slightly grooved; this is slipped on to the front of a fully corrected lens in the same manner as a filter.

CHAPTER III

THE MECHANISM OF THE CAMERA

Principles of design

A camera is a light-tight box holding the sensitive material, a lens or other means of focusing the image, and, usually, a shutter to control the time of exposure. To this is generally added some

Fig. 33. Fixed focus box camera.

arrangement for viewing the objects included on the negative. The simplest form of camera is a rigid box (fig. 33). It is not strictly necessary that the camera should have a lens; this may be replaced by a pinhole. The pinhole camera is a direct descendant of the old camera obscura. Occasionally a pinhole camera has advantages over the ordinary one equipped with a lens of the conventional angle of view, e.g. for photographing buildings when it is not possible to

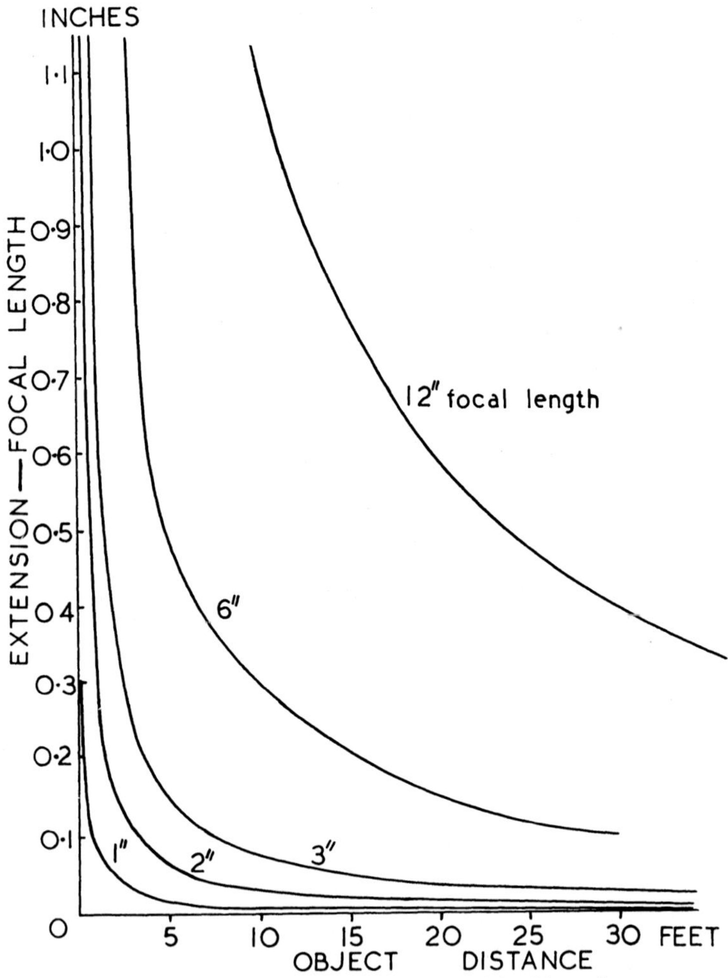

Fig. 34. Variation of camera extension with object distance.

get to any great distance from them. The best distance from pinhole to plate is found to lie between the limits calculated by multiplying the square of the diameter of the pinhole by a factor between 625 and 1250.

Before discussing cameras in detail there are certain principles to be taken into account. First, it is necessary to fix some sort of relation between the focal length of the lens and the size of the negative, otherwise correct perspective may not be obtained. The conventional relation is that the focal length should be approximately equal to the diagonal of the negative. The question of perspective is dealt with on p. 91. As was shown in Chapter II, the simple equation

$$\frac{1}{i}=\frac{1}{f}-\frac{1}{o}$$

gives the relation between distance from lens to image, i, as a function of the distance of object, o, and focal length, f. Fig. 34 shows the necessary alteration of extension, i, for infinity to 3 feet. At infinity, from the equation, the distance from lens to plate is equal to the focal length of the lens. Fig. 34 therefore shows the extra extension beyond this required for closer object distances. It is seen that the extension necessary becomes quite small when the focal length is 3 inches or less and also that the extension is very small for such lenses between infinity and about 8 feet. It is therefore possible to make box cameras with fixed focus inside these limits. It also follows from this figure that miniature cameras need very small total extension and, conversely, that the accuracy of focusing must be correspondingly greater. The modern miniature camera is a box camera but not a fixed focus one (fig. 41). This form is obviously desirable for cameras with short focal-length lenses, since any folding camera is less rigid. It may be noted that a fixed focus box camera, at its proper distance, gives images as sharp as those from any other camera with the same lens.

Methods of focusing. Field cameras

The types of camera are best classified according to their focusing devices. The commonest uses a scale of object distances attached to the base board to give correct extensions. This is quite satisfactory for lenses of focal length greater than about 5 inches. The oldest type of camera was fitted with a piece of ground glass in the focal plane; the subject was focused visually by rack and pinion movement of the lens. Fig. 35 shows a modern form of this type which is usually called a 'field camera'. This method is sound, but has the drawbacks that the image is inverted and the photographer needs to provide himself with a black cloth over his head to cut out extraneous light.

Fig. 35. Field camera with double extension—first by rack and pinion, second by front struts; lens panel rising in front which can also be tilted by the struts; tilting back with vertical position indicator; the back can be racked forward for wide-angle lens work. A focal-plane shutter is fitted.

Reflex cameras

The next type of camera to appear was the reflex (fig. 36); here the image is reflected from a silver surface mirror, *M*, on to a ground-glass screen, *G*, which is inspected from above. A hood, *H*, cuts out extraneous light. Reflection by the mirror puts the image the right way up again. This type of camera is the best, taken all round, for serious work although, by modern standards, it has the disadvantage of being rather bulky. It is obvious that the box must be at least as deep as the focal length of the lens. Owing to the method of focusing, a focal plane shutter, *S*, is usually fitted. The reflex camera then has the advantage of easy interchangeability of lenses. In any type of ground-glass screen focusing extra accuracy can be obtained by the parallax method.

A pencil mark is made on the ground glass and, on it, a microscope cover glass is cemented with Canada balsam. This area is examined by a magnifying glass. When this is adjusted to give a sharp image of the pencil mark, and the camera is focused so that the image can be seen also, any displacement of the eye will cause

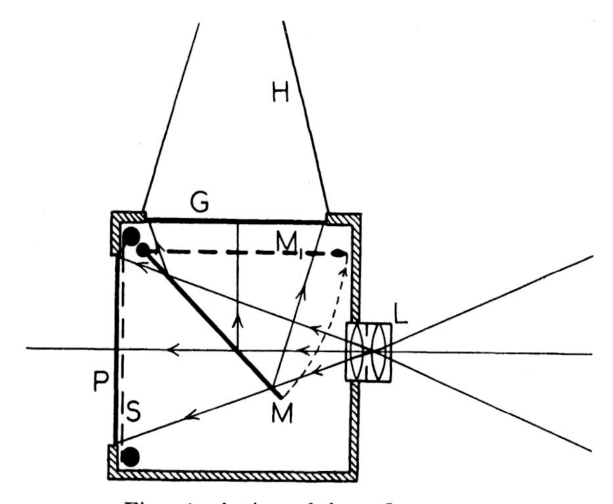

Fig. 36. Action of the reflex camera.

no movement of the two images relative to one another. When focusing is not correct, relative movement occurs; if the pencil mark is displaced, relatively to the image, in the same direction as the movement of the eye, then the lens extension is too great. If the pencil mark moves below the image as the eye is raised, then the extension is too short.

The ground-glass screen must be fairly coarse to scatter the light at the edges of the plate. For fine focusing, however, it should be more finely ground in the centre. Alternatively, a little vaseline may be rubbed on to the centre part.

Coupled range-finders

Recently a new method of focusing has been applied to cameras. A range-finder of the military type is fitted to the camera and coupled mechanically to the lens-focusing movement. In some makes of camera, focusing the lens moves the range-finder control, while, in others, a single control adjusts the two movements simultaneously. In both cases, sharp focusing is shown by the point at which the double image, seen in the finder, becomes single. The base line of any camera distance meter is automatically small owing to the small size of the camera into which it is built. An interesting account of coupled range-finders is given in the

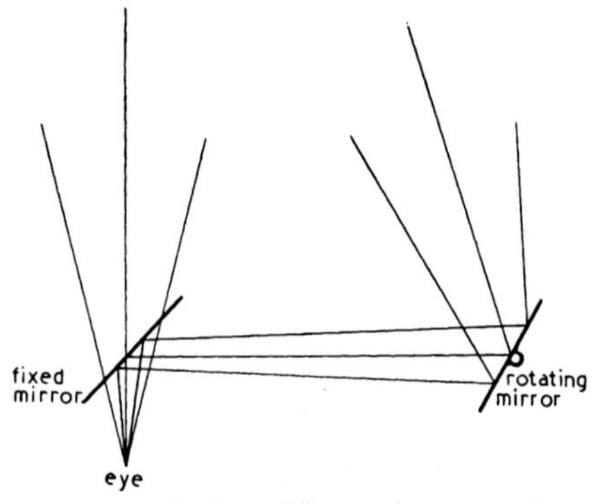

Fig. 37. Range-finder with two images coincident.

American Annual of Photography 1939, p. 55, by I. Clyde Cornog. The theoretical limitations are discussed for a number of types. Fig. 37 shows the principle of the simple range-finder.

The lens

The choice of a lens is usually dictated by price. Except for speed work there is no particular advantage in using a lens of very large aperture (see p. 44). For special subjects it may be desirable to have a telephoto and a wide-angle lens. For general purposes, however, it is very much better to buy a camera, in the first place, with a very wide range of extension. The Sanderson camera (fig. 35) is a good example of this type; it has a double extension for very close-up subjects where a large scale of reproduction is required and it can also be used with the necessary small extension for a wide-angle lens. The extension for wide-angle lenses is, of course, much smaller than that required for the normal lens whose focal length is equal to the diagonal of the plate. In using a wide-angle lens with small distance between plate and lens, the front of the base plate of the camera may be in the field of view. In some makes of field cameras this is avoided by having it hinged.

In others, the lens panel is fixed, and the plate and the back of the camera moved forward by rack and pinion. For record work it is often convenient to attach the lens to an extension tube, which is screwed into the front panel of the camera. In this way extensions necessary for reproduction at actual size or larger can be given. Under these conditions the depth of focus becomes very small and the lens has to be used at a small stop. Care must be taken with extension tubes to see that no part of the image is cut off by the tube.

The diaphragm

The first form of diaphragm consisted of metal plates, with holes of various size in the middle, which were dropped into slots in the lens mount. These were displaced later by a single metal plate, with a number of holes of different sizes, which was rotated till the desired one was in position. These diaphragms have all been superseded by the iris diaphragm. The aperture in this is varied by rotating the outer casing. It may therefore be built into the lens mount.

The iris diaphragm, made of pivoted plates, should be treated with great care and no attempts at cleaning or adjusting should be made except by an experienced worker. On no account should oil be applied to the leaves. This also applies to shutters of the diaphragm type.

The shutter

The shutters used are of two main types: focal-plane, which work by some sort of a blind moving across just in front of the plate; and those which work just behind the lens, between its components, or in front of it. The focal-plane shutter consists of two blinds. The exposure is varied by altering the width of the slit and sometimes also by the tension on the blind. The focal-plane shutter is convenient, but occasionally causes a peculiar type of distortion. When an object is moving at high speed at right angles to the direction of travel of the slit, it may have moved quite a considerable distance during the time which the slit takes to move from top to bottom of the plate. Roller-blind shutters, made as separate units, can be fitted on the front of the lens. The between-lens shutter consists of pivoted plates, similar to those in the iris diaphragm, worked by a spring. Considerable caution should be used in trusting the speeds marked on these shutters.

For slow speeds, such as are used in portraiture, a very convenient form of shutter is a double flap worked by air pressure from a rubber bulb.

The release of the shutter is a matter of some importance. It is quite easy with most cameras to press the release without moving the camera. It is, however, also very easy to shake the camera badly, and special releases, consisting of Bowden wire cables, are made to avoid this. Some of those on the market, however, are so short that a considerable amount of vibration may be transmitted along them.

Extra movements and attachments

For a general utility camera, extra movements are necessary. The most important of these is the rising front. The usual form allows the lens panel to be moved up and down by rack and pinion. This type of movement is essential for photographing high buildings or similar objects where the position of the camera is much below the centre of the object. For this type of subject a tilting back and rising front are also useful. When using these movements the back must be kept vertical, also the lens panel, the position of the lens being chosen as desired. Fig. 38 (*a*) shows what happens if the camera is tilted to include the top of a high building, or for any other reason. The rays from the top part do not have so far to travel before they reach the plate, so that dimensions of the image are proportionately reduced. This gives rise to the appearance of converging verticals so frequently seen in photographs of buildings. Conversely, if the camera is tilted downwards, the verticals will diverge upwards. These effects can be avoided by tilting the plate holder until it is parallel with the object; that is, perpendicular, if the object is also perpendicular. The lens, however, must be stopped down considerably, as a large depth of focus is required for both top and bottom to be in sharp focus. Alternatively, the lens may be raised and the camera kept horizontal (fig. 38 (*b*)). The most satisfactory method for extreme cases is to use all three methods together. The axial rays can then be arranged to fall on the centre of the plate (fig. 38 (*c*)).

It should be noted that this bugbear of tilted verticals is always present and care must be taken to keep the camera horizontal. Field cameras are usually provided with a spirit-level to check this

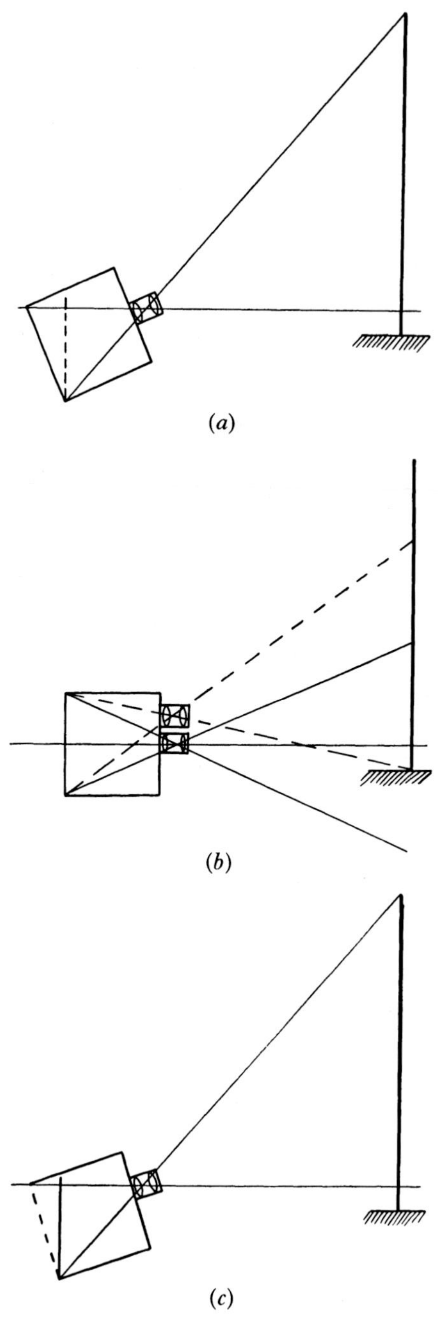

Fig. 38. Rising front and tilting back.

point, but with ordinary hand cameras, it is a matter of guesswork. With a reflex, a convenient guide consists of a pencilled cross on the centre of the ground glass. A convenient object is selected at the same height as the camera lens (not the operator's eye level) and the camera held so that its image lies on the cross on the

Fig. 39. Contra-jour lighting using lens hood.

ground glass. Tilted verticals may be corrected in enlarging by tilting the plate in the enlarger and the paper on the easel so that the original tilt is compensated. Pictorially, slightly tilted verticals are most objectionable, but great convergence may be used to give a powerful impression of height. For record work no distortion is permissible.

When the camera is used on a tripod, attention should be given to the tripod bush on the camera, as these are often quite inadequately fixed. Tripods themselves must also be regarded with suspicion, since the lighter ones are often unstable. When using a reflex camera, instability may be caused at speeds less than $\frac{1}{50}$th of a second by the rather heavy action of the mirror release. In case of doubt it is better to stop down the lens and give a much longer exposure without using the releases. This can be done conveniently by means of a loose cap over the lens, the camera having been set for a time exposure.

An indispensable accessory to every lens is a hood to cut out extraneous light, which causes veil and consequent diminution of the quality of the negative. The lens hood should be constructed with a slot into which filters can be slipped just in front of the lens. Fig. 39 shows an example of a picture taken against the light with a hood. Such pictures are impossible without a hood because the curved front glass of the lens catches the light, which reaches the negative by internal reflection, refraction and scattering. A more or less diffuse patch of light will appear on the print in consequence.

Choice of camera

The choice of a camera obviously depends upon the work for which it is intended. In general it is best to keep the size small for convenience, but not so small that any loss of quality is seen when enlarged pictures are made. Also not so small that the negatives need special processing with fine-grain developers. This sets a lower limit of $4\frac{1}{2} \times 6$ cm., while general convenience sets the limit at half-plate, $6\frac{1}{2} \times 4\frac{1}{4}$ inches, for maximum size. In the last chapter the additional advantage of depth of focus in the smaller camera has been described. They are also cheaper in cost of negative material although, if reasonable intelligence is used, this last factor is not serious. The difference, however, between storing big negatives and small ones may be very serious. For laboratory work, where lantern slides may be wanted, it is desirable that they should be made by contact printing. This means that the camera should be a quarter-plate one or $2\frac{1}{4} \times 3\frac{1}{4}$ inches.

Focusing is almost always important in scientific record work. Whether a reflex or a ground-glass focusing screen or a coupled range-finder is used must be determined by the nature of the work to be done. If interchangeability of lenses is not wanted, then the coupled range-finder type is the best. Fig. 40 shows the Zeiss super-Ikonta, which is, in my opinion, the finest camera for the amateur on the market. Fig. 41 shows the Leica 35 mm. film camera. This is the first of the precision miniature cameras to be put on the market and is particularly useful where photographs have to be taken quickly or under awkward conditions.

The greatest credit is due to Messrs Leitz for having taken the commercial risk of introducing the miniature camera, which had to be a scientific instrument made by precision methods and which

Fig. 40. Zeiss super-Ikonta camera with range-finder coupled with focussing and combined with view-finder.

therefore had a correspondingly high price. The greater range of optical equipment possible with the small focal length required by the miniature negative, especially the large aperture lenses which can be used with it and which could not be made or used with large camera sizes, requires the most exacting standards of design, workmanship and quality of negative material. All these are now provided by the leading manufacturers and cameras of larger sizes, especially the $4{\cdot}5 \times 6$ cm. and $3\frac{1}{4} \times 2\frac{1}{4}$ inches, have gained by generally improved standard of design and construction and by inclusion of additional devices, such as range-finders, originally introduced for the miniature camera using 35 mm. negative.

When plate cameras are to be used for film, there are three possibilities. A film pack is perhaps the simplest, but these are much more expensive than roll film or cut film. A film-pack adapter is necessary in any case. Roll-film adapters can also be bought or can be made quite easily in the workshop. Cut film is the most convenient for a number of purposes. Its lightness

Fig. 41. Leica "miniature" camera.

compared with plates becomes important when large numbers of negatives have to be stored. Its rigidity compared with the thinner film used in roll-film packs is useful in larger sizes. All that is required for its use is metal sheaths, into which the cut film slips, which are then inserted in the dark slide as if they were plates. These sheaths hold the film flat and bring the total thickness up to that of the glass plate.

Roll-film cameras need a window at the back for observing the numbers when winding on the film. In all cheaper cameras these are red. This window is not safe against direct sunlight with panchromatic films.[1] A small gadget (supplied with each roll

[1] A sample of red-window material was tried by the writer as a filter. With panchromatic material it acted as a very good red filter requiring about 12 times exposure!

film) can be attached as a safety cover and more expensive cameras are now fitted with a built-in sliding metal one. With cameras in which there is a film counter, it is only necessary to use the window to wind to Number 1 exposure. This can be done in a dim light and no further care need be taken provided the camera is not exposed to bright sunlight. In an ever-ready case this, of course, never happens.

The choice between film and plate is usually settled by the apparatus in use. Otherwise the choice is best settled by the frequency with which photographs are to be taken. If frequently, film is more convenient, but when occasional ones only are taken, plates (or cut film in the plate holders) must be used. For larger sizes of negative, such as half-plate, glass plates should be used to ensure flatness of the negative. In smaller sizes the arrangements in all properly made cameras are sufficient to ensure the film being flat. Plates, especially the larger sizes, become intolerably bulky in storage. This difficulty is partly evaded by cut film. It can be trusted to lie flat, since the metal sheaths which fit into the ordinary plate holders keep it flat within the limits required. Cut-film base is considerably thicker than that used for roll films, so that there is no difficulty in development in dishes. Halation is, of course, greater in the thicker cut film than in roll film or film pack, but less than in plates (p. 27).

Care of camera

Finally, remember that a camera is a precision instrument. This does not mean that it is so well constructed that it may be abused. Quite the reverse. It means that it will only function at full efficiency so long as it remains a precision instrument. It should always be kept in a case except when in use. Smaller cameras can be kept in an ever-ready case, which has a flap on the front which lets down for use. Every time that the camera is opened for putting in a new film, it should be cleaned from any accumulated dust or wood shavings. In dusty or sandy places, the finer particles penetrate through all the joints with the greatest ease. These can be removed when the camera is opened, but it is better to prevent their entry as far as possible by keeping the camera in a good case.

Cameras with leather bellows may develop light leaks which lead to a peculiar fault in negatives. The hole acts as in a pinhole camera and can form an image of the sun provided the direction is right. As the camera is moved about slightly, the image will be drawn out into a wavy line. Such pinholes are easily detected by taking the camera into a dark room and placing a lamp inside. The light coming out through leaks is then easily observed.

Other faults likely to develop in old cameras, or in cameras which have been badly handled, are in the shutter and, in reflexes, tarnishing of the mirror. Both of these are jobs for experienced repairers. It may be noted here that shutter speeds should never be trusted for accurate work. If comparisons need to be made, alteration of exposure should be made by altering the stop, keeping to one marked shutter speed. It is hardly necessary to add that the lens must always be treated with great care. Cleaning should be done only with a cloth free from any grit or dust. Bubbles in the lens glass are of no significance to the sharpness of the image. They are often unavoidably present in some of the glasses needed for lens components. Occasionally a coloured film appears on the front surface of the lens. This is not serious and the owner should on no account try to remove it.

Ciné cameras

Ciné photography is frequently of value, both in making a permanent record of changes not otherwise easily recorded, and for analysing rapid changes. The basic principle is that films should be moved rapidly past the focal plane, being held stationary for the duration of the exposure. For ordinary speeds this can be achieved easily by mechanical means, but for very high-speed work the film cannot be stopped and started intermittently and optical methods have to be employed. In viewing the film, this intermittent movement is seen as a continuous one, owing to the physiological phenomenon of 'persistence of vision' which occurs when the pictures are shown at not less than 16 per second.

Fig. 42 shows the principle of the ciné camera. The winding mechanism (not shown) turns the film spools, the two sprockets, *S*, which engage with the holes punched along the sides of the film, and the crank mechanism of the claw which gives the film its discontinuous motion. Since the rest of the winding is continuous, a loop of film is left between each sprocket and the claw. Exposure is made by the rotating shutter, which is a disc of metal with a slot in it. The mechanism is adjusted so that the shutter slot is opposite the lens and allows the light to pass through on to the film when this is stationary. While the claw is pulling the film past the focal plane, the light is cut off.

Ciné film is made in three sizes; full size, 35 mm., used by cinemas, 16 mm. and 8 mm. The two latter are suitable for amateur

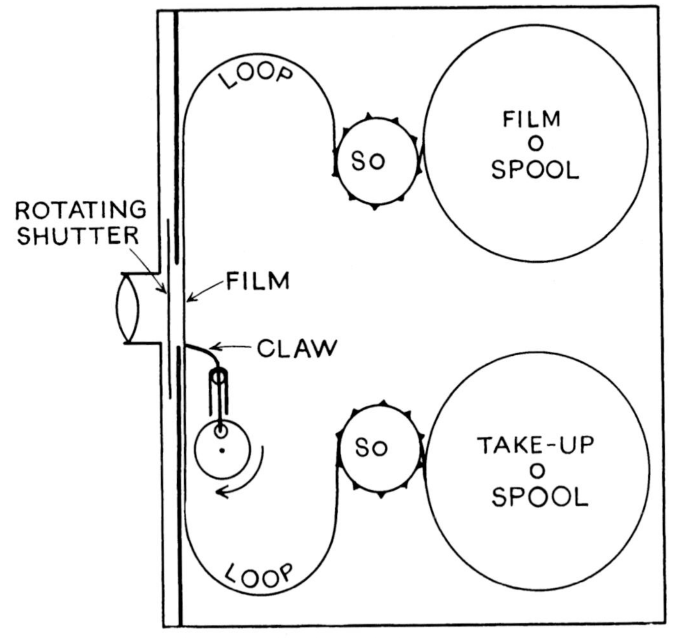

Fig. 42. Working of ciné camera.

work. Cameras can be obtained reasonably cheaply and, especially for the 8 mm., the optical equipment is very simple. The small focal length of the lens gives a very big depth of focus without any focusing even at large apertures. Exposure is usually controlled by the iris diaphragm. With the more expensive cameras the speed is variable and sometimes an adjustment can be made to take single exposures. This is particularly useful in obtaining records of slow changes, which can then be speeded up by projection of the film. Fig. 71 (*b*) shows every hundredth frame from such a film record of the division of sea urchin eggs after fertilisation. Fig. 71 (*a*) shows another example of scientific use of the ciné camera.

Processing of the film is usually included in the price of the stock. In the smaller sizes the film is usually reversed after development to give a positive.

Some modern ciné cameras are being fitted with built-in exposure meters which are coupled to the diaphragm. This method of automatic exposure control is obviously much simpler mechanically than alteration of the exposure time by the meter. This latter method, however, is the preferable one. Alteration of the aperture is the means of controlling depth of focus. For pictorial work it is essential that this should be free. It is true that in ciné cameras, especially those of small size, the depth of focus is very great under all conditions, so that loss of control of it is not so important. However, it is most undesirable that this feature should be applied to larger cameras.

16 mm. and 35 mm. films are now being used in rapidly increasing amount for recording purposes. The *New York Times* had ninety thousand pages, covering the years 1914–1918, copied. The total cost was about £90. Tests suggest that the films will be as durable as the original paper. They have the great advantage, however, that they can be copied easily at any time. Banks are using 16 mm. film for photographing cheques; the Kodak Recordak apparatus can record 5000 per hour. The machine is automatic and requires no photographic experience to use it. It is believed by a number of people that the time is rapidly approaching when a film service will be an integral part of every big library, especially technical ones. Anyone requiring information will then be supplied with film copies of anything that he requires. High speed ciné work is described in Chapter v.

The Board of Education has published a pamphlet, (1938) 1/6, No. 115, 'Optical Aids', on projectors for schools. This deals with both ciné and still projections. It is characteristic of the present attitude of those in authority to the question of education by the film that no mention is made of Universities.

The number of documentary films made is larger than the number of entertainment ones. The number, however, with real scientific value is minute. The reason is quite simple. Some one must pay for the cost of making the film. If a commercial concern is called in to make the film its charges will be very large and the interference with normal routine may be serious. There is, however, no reason why numerous demonstrations, too complicated to be carried out at lectures, should not be filmed and shown at every University in the world every year, except that the authorities concerned do not realize the power of the film for education and do not know that making such a record film would be quite a simple matter in any laboratory possessing staff with quite elementary technical knowledge. There are still people who object to the showing of technical films at lectures because it makes them too like entertainments.

CHAPTER IV

COLOUR PHOTOGRAPHY

The three-colour principle

The principles of colour photography date back to 1861, when Clerk Maxwell showed that all colours may be produced by mixing, in suitable proportions, three primary colours—red, green and blue violet.[1] Maxwell took three photographs, one through a

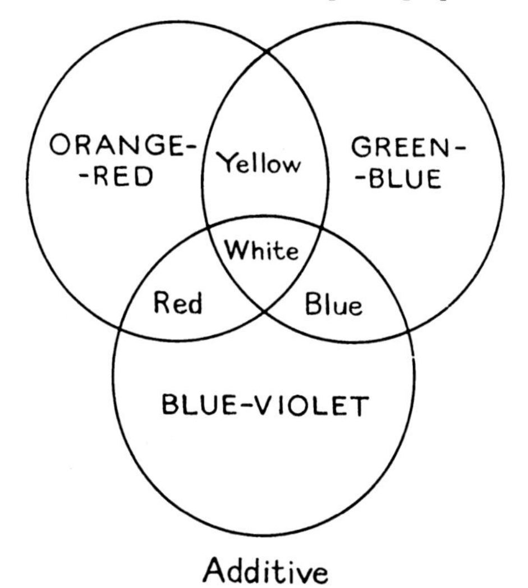

Fig. 43. Colour mixing.

red filter, one through a green and one through a blue-violet; from these he made positive lantern slides. These were projected by three lanterns, each through the filter originally used. Mixture of the three colours in equivalent intensity produces white light: red and green make yellow, violet and green make blue (fig. 43). This is the principle of additive colour methods but, in practice, they have not been successful owing to the difficulty of superposing the three images except in the case of projection by three separate lanterns. The alternative method, which has proved much more amenable commercially, is the subtractive one. It is possible to find dyes

[1] Suggested originally by Thomas Young.

which are not pure colours themselves but minus pure colour; thus we can have a yellow which will be minus violet, a pinkish-red which is minus green, and a bright blue which is minus violet. Superposition of these three minus colours cuts off everything and therefore appears black; the blue and the pink give violet, the blue and the yellow give green (fig. 44).

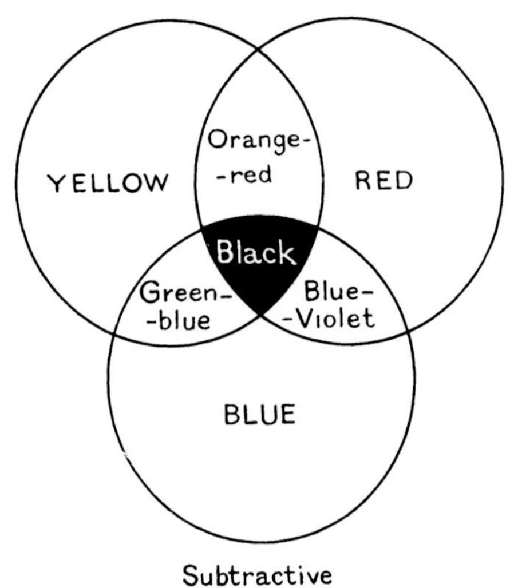

Subtractive

Fig. 44. Colour mixing.

Additive processes give poor light transmission and brilliance compared with subtractive for the following reason. Any colour is produced from the white viewing light by filtration of the complementary colours. White light is then produced by mixture of the three colours after removing their complementaries. The total white is therefore small compared with a directly transmitted white. In the Dufay process, the colour filtration is not done as such, but by the equivalent method of dividing the total area between the three colours.

The development of methods of colour photography

In 1891 Lippmann devised a method of colour photography which gives results of remarkable perfection. A panchromatic negative as grainless and transparent as possible is exposed through the glass, the emulsion being in contact with a bath of mercury,

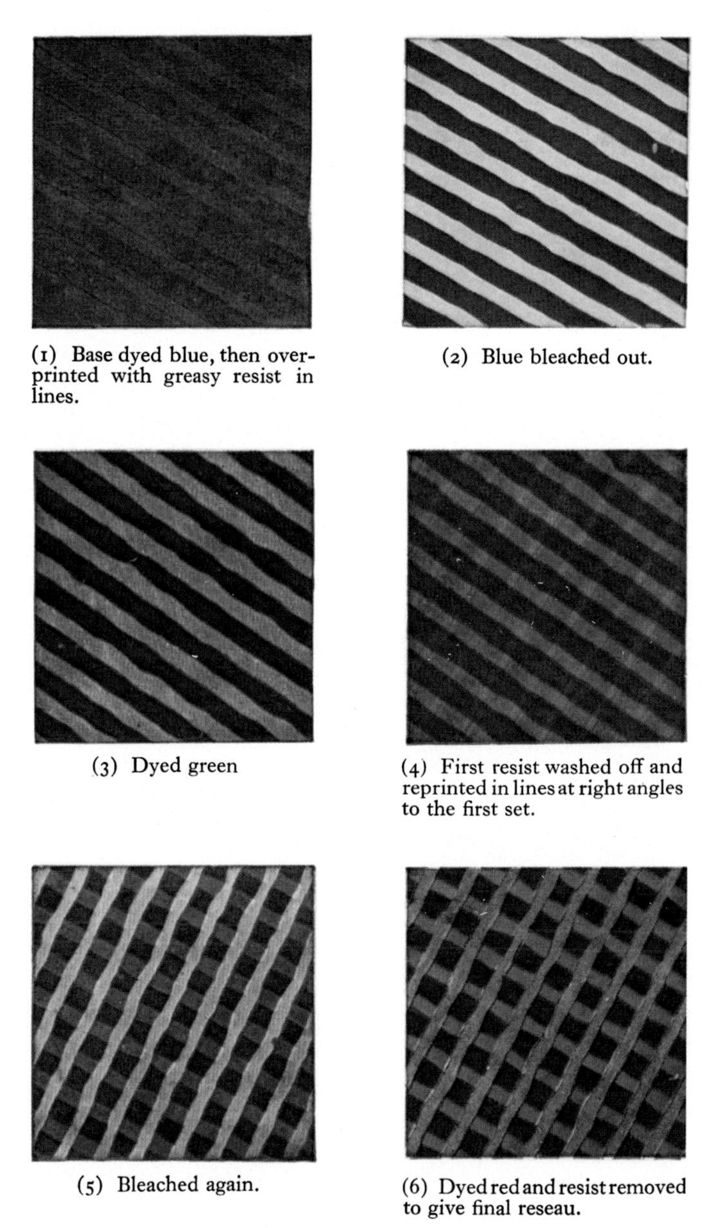

(1) Base dyed blue, then overprinted with greasy resist in lines.

(2) Blue bleached out.

(3) Dyed green

(4) First resist washed off and reprinted in lines at right angles to the first set.

(5) Bleached again.

(6) Dyed red and resist removed to give final reseau.

These photomicrographs were taken on Dufaycolor Film, the reproductions being made from the transparencies.

Magnification × 100 Linear.

Fig. 45. Photomicrographs of Dufaycolor Reseau.

which forms a mirror in optical contact with the emulsion. The light passing through the emulsion is reflected at the mercury surface and interference bands are formed in the emulsion. On development, silver is deposited along the bands, which are automatically spaced at the correct distance to give interference colours corresponding to the colour of the incident light. This method, however, is not possible commercially.

Colour separation has been achieved by embossing the surface of the film with tiny cylindrical lenses which separate the spectral colours. After exposure the film is developed, reversed and projected by an optical system which includes a filter containing narrow bands of the three primary colours arranged to correspond with the colour patches on the film. This was the basis of the Kodacolor process (now obsolete).

Screen methods

For still photography a number of methods have been devised in which some sort of a screen is used. In 1894 Joly made a filter by ruling alternate lines of red, green and blue-violet on a glass plate (80 of each colour per inch) through which a negative was made. The positive was then viewed through a similar screen placed in correct register. This was improved by Finlay, whose filter has a chequer pattern of the three colours, a similar screen being bound up with each positive.

Meanwhile the mechanical difficulty of producing a three-colour separation filter of sufficient fineness was evaded by the ingenious Lumière autochrome process. Three separate lots of starch grains are dyed the three colours, mixed together, coated on a glass plate and the emulsion superposed. Exposure is made through the glass and filter; the plate is developed and reversed and a positive colour transparency obtained. A modification of this process, made by Agfa, consisted of using grains of shellac in place of starch. One of the most serious objections to these processes is that, although the grains may be adequately small individually, mixing can never produce a perfectly homogeneous mixture and clumping of several grains of one colour is liable to occur so that the final filter is coarser than the dimensions of the single grains suggest. The non-uniform colour sensitivity of the emulsion may be corrected by varying the proportion of grains of the three colours.

The mechanically produced filter has recently been very greatly improved in the Dufaycolor method. Fig. 45 shows the appearance of the filter at its various stages of preparation. The supporting film base is coated with a thin layer of collodion dyed blue; lines are then printed in greasy ink across this; the dye not protected by the ink line is bleached out, and the film is then washed and dyed green. The first set of ink lines are now removed and a second set printed at right angles to the first. Another bleaching bath removes green and blue dyes between the second set of lines. In the last bath these clear spaces are dyed red and the ink lines removed. The red lines are narrower, so that areas of the three colours are obtained in the final screen to suit the colour sensitivity of the emulsion used. After protecting the filter screen with a thin layer of varnish, panchromatic emulsion is coated on top. The film is exposed through base and screen, developed, reversed and viewed as a positive transparency. Fig. 46 shows an example.

Dufaycolor is thus an additive system with considerable practical advantages. No special apparatus is needed either in taking or in processing. Being additive, it has, however, the serious disadvantage of reduced transparency. The intensity of light passing through it is small; it is obvious that if we consider one colour, say red, only about one-third of the film area of a red image will be transmitting light. This, coupled with the finite dimensions of the screen, limits the use of the Dufaycolor method seriously for large-scale cinema work. It is, however, a very valuable method for still photography for amateurs. Incidentally, it has the advantage that under-exposure intensifies the brilliance of the colours, provided, of course, that the under-exposure is not too great. Over-exposure gives washed-out colours, since any colour becomes progressively mixed with the complementary ones as exposure is prolonged.

Dye-coupling developer methods

An entirely new approach to the problem of colour photography has become possible by application of the method of dye-coupling to produce the colours. Instead of using incorporated colours which act, by the various means already described, as filters, there are applied either in a processing bath or by incorporation in the emulsions, substances which combine with the oxidation product of the

Fig. 46. Reproduction of Dufaycolor picture.

developer and form coloured substance. The reaction in a simple case is as follows:

paraphenylene diamine + silver halide + phenol → blue dye + silver

$$H_2N-C_6H_4-NH_2 + 3AgCl + C_6H_5-OH \rightarrow$$

$$H_2N-C_6H_4-N=C_6H_4-OH + 3Ag + 3HCl$$

The hydrochloric acid formed is neutralized by the sodium carbonate in the developer.

Suitable dyes for covering the spectrum have been found and grainless images are achieved by placing the three emulsions which are to form the three colours on top of each other. Suitable colouring matter beneath the top layer prevents light of the colour which forms its latent image from reaching the lower layers. The two remaining colours can be separated by using layers of different colour sensitivity.

The two commercial applications of this method of colour photography are Kodachrome and Agfacolor. These differ in that separation of coloured images is achieved by the former by controlled diffusion during processing, whereas the latter uses the simpler method of forming the required colour from the components *in situ*. These processes are grainless ones, since their action does not depend upon a ruled screen and the images contain no silver. The separation of the three-colour images is achieved not by lateral separation, but by using three separate layers of emulsion placed on top of each other. Fig. 47 shows a cross-section of Kodachrome film. It may be noted that this is a subtractive process. To make a colour picture, the three images are transformed into three-colour positive images; that in the red-sensitive layer becomes the blue-green image; that in the green-sensitive layer becomes the magenta positive, while the top blue-sensitive layer yields the yellow positive. After exposure the images are first developed to form an ordinary negative; the negative image is removed in a reversing bath, and the film exposed to light. It is then redeveloped with a coupling developer; that is, one in which the oxidation product of the developing agent combines with a suitable substance in the solution to form an insoluble dye. At this stage the film, therefore, contains blue dye and silver. The film is then fixed, washed and dried. The silver is now reconverted into silver halide by a bath which destroys the dye in the top two layers only. After washing, the film is treated with a second coupling developer which coats the top two layers magenta. The film is washed and dried again, and then bleached, thereby destroying silver and red dye of the top layer only. The last stage consists of developing with a third coupling developer which deposits silver in the top layer and dyes it yellow. Now the silver throughout the film is converted into halide, which is removed by a fixing bath which does not affect the three dyes. The final film therefore contains the three dye images and no silver. It is, therefore, practically grainless. This method is obviously of the greatest value for cinema work, owing to high transparency and freedom from grain. It has the drawback, however, that the processing requires special plant because of the need for and difficulty of controlling the three coupling-development processes. In general, for accuracy of colour reproduction, Dufaycolor has the advantage that any dye may be chosen, whereas in the Kodachrome process only those formed by coupling developers can be used. It should be noted, however, that no process involving the use of dyes can reproduce spectral colours purer than the dyes themselves. For this reason a spectrum cannot, therefore, be accurately photographed by these methods. Fig. 48 shows an example of Kodachrome.

The latest Agfacolor process is a subtractive one in which three emulsions 0·005 mm. thick, separated by layers of gelatine 0·002 mm. thick, are coated on the film base. The top emulsion is blue sensitive and beneath it is a yellow filter layer. The middle emulsion is green sensitive and the lowest red sensitive. After exposure, the three latent images are developed in an ordinary developer. The film is then exposed to light and redeveloped in a paraphenylene developer. The originally unexposed silver is now reduced, and at the same time a dye image is formed by coupling of the oxidation products with substances present in the emulsions. Processing is then completed by treatment with a mild oxidizing solution which removes the silver.

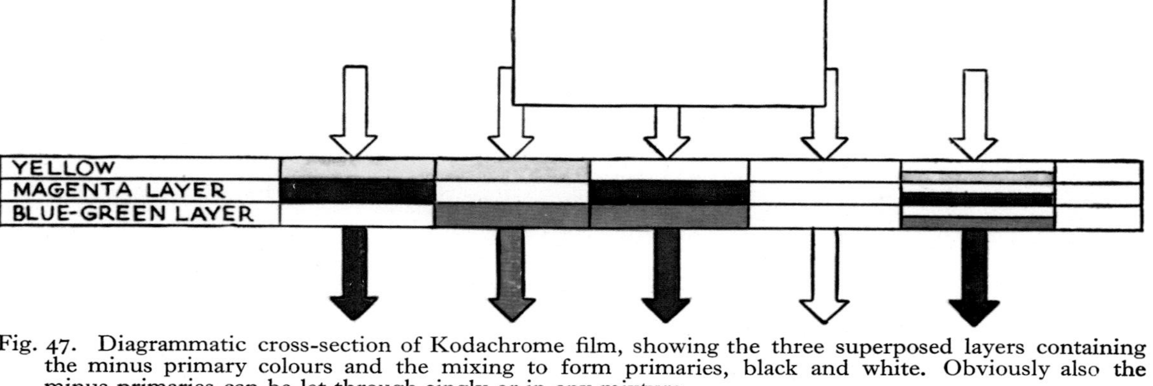

Fig. 47. Diagrammatic cross-section of Kodachrome film, showing the three superposed layers containing the minus primary colours and the mixing to form primaries, black and white. Obviously also the minus primaries can be let through singly or in any mixture.

The use of colour photography

Since the introduction of panchromatic emulsions and suitable correcting filters, it has become common for people to speak of 'correct rendering of colour in monochrome'. What they really mean is so obvious that it is easy to overlook the basic absurdity of the claim. Colour is colour and cannot be rendered in monochrome. It is true that we can reproduce a series of tones which correspond to the visual luminosities of colours, but that is quite a different thing. Consider the copying in monochrome of a green and red pattern. With suitable shades, the visual luminosities will be equal and no pattern will be seen at all, and the 'correct' rendering in monochrome of the pattern is a blank grey. Alternatively, if one wants to show the design, the colours must be reproduced incorrectly. Although this is a somewhat extreme case, the general principle holds good for photography of all coloured objects. The use of correcting or distorting filters depends on this principle and is often of the greatest value. Decide what objects are to be specially emphasized. Decide whether you want them emphasized by being light on a dark background or dark on a light one and then choose your filter so that, in the first case, it transmits the colours of the objects and not of the background; and, in the second, the converse. Correct reproduction often leads, especially in landscape work, to undue flatness because everything except white clouds is rendered in middle grey tones. This does not occur so much in other pictorial work, since there the necessary dark and light tones are produced by the lighting and shadows.

This question of colour spoils so many landscape pictures which have obviously been taken because their colouring impressed the photographer. It is essential to realize that any coloured scene may be reproduced in colour according to two quite separate and contrary styles. Objects of interest may be built into a composition, helped by the lighting and shadows and then still further helped by colour differentiation. Against this, there is the method of using colour contrast alone to give the impression which struck the observer. In the extreme cases of some of the impressionists, the ordinary devices of composition, even such as perspective, are discarded, all composition being due to colour differentiation. Which method an artist uses depends upon himself, and upon which happens to be the conventional one at the time when he works. The majority use a combination of the two. The extent to which the artist tends to one method or the other is fixed to some extent by the nature of the subject. Landscape can be done by either alone or by any combination of the two. A subject, such as a nude, generally leads the treatment automatically to the compositional side because a particular nude is being copied. Of course, this need not be done, and an impressionistic idea may be put over rather than a picture of a given nude.

This discussion is of special importance to the photographer using colour methods, since he is, by training, always trying to copy something; to make a picture of things that he sees as he sees them. The result is that he tends to use colour only as an aid to his composition, and never to use colour alone; thus we get the familiar colour transparencies full of blobs of colour instead of real compositions of colours, sheets of colours themselves forming a design. Of course the extreme case of this is the abstract design in colour.[1] This has been done already by photographers, but it is going rather far for most. The general rule of looking for large masses of colours rather than looking for fields of tulips and the like is a good rule. Blue-violet light scattered from atmospheric haze or from excess from clear sky in the summer (cf. fig. 13) may spoil colour photographs if not eliminated by a suitable filter.

Colour photography is used but little by the scientific worker, probably mainly owing to the difficulty of assessing the correct exposure. This difficulty is not so serious as it appears, since trials may be made with ordinary plates rated by the manufacturers as of the same speed. The difficulty already mentioned, of incorrect colour rendering of colours purer than those used in the process, is serious in certain types of work. Nevertheless, a slightly inaccurate colour record will in most cases be more instructive than a description or a black and white rendering. Colour records of great medical value have been made already. When, if ever, records will be used systematically for instructing students remains to be seen. The cost of reproducing colour pictures in books and journals is a serious deterrent, but the cost of making copies of transparencies for projection at lectures is trifling.

[1] The brilliant work of Len Lye in films made for the General Post Office Film Unit should be mentioned.

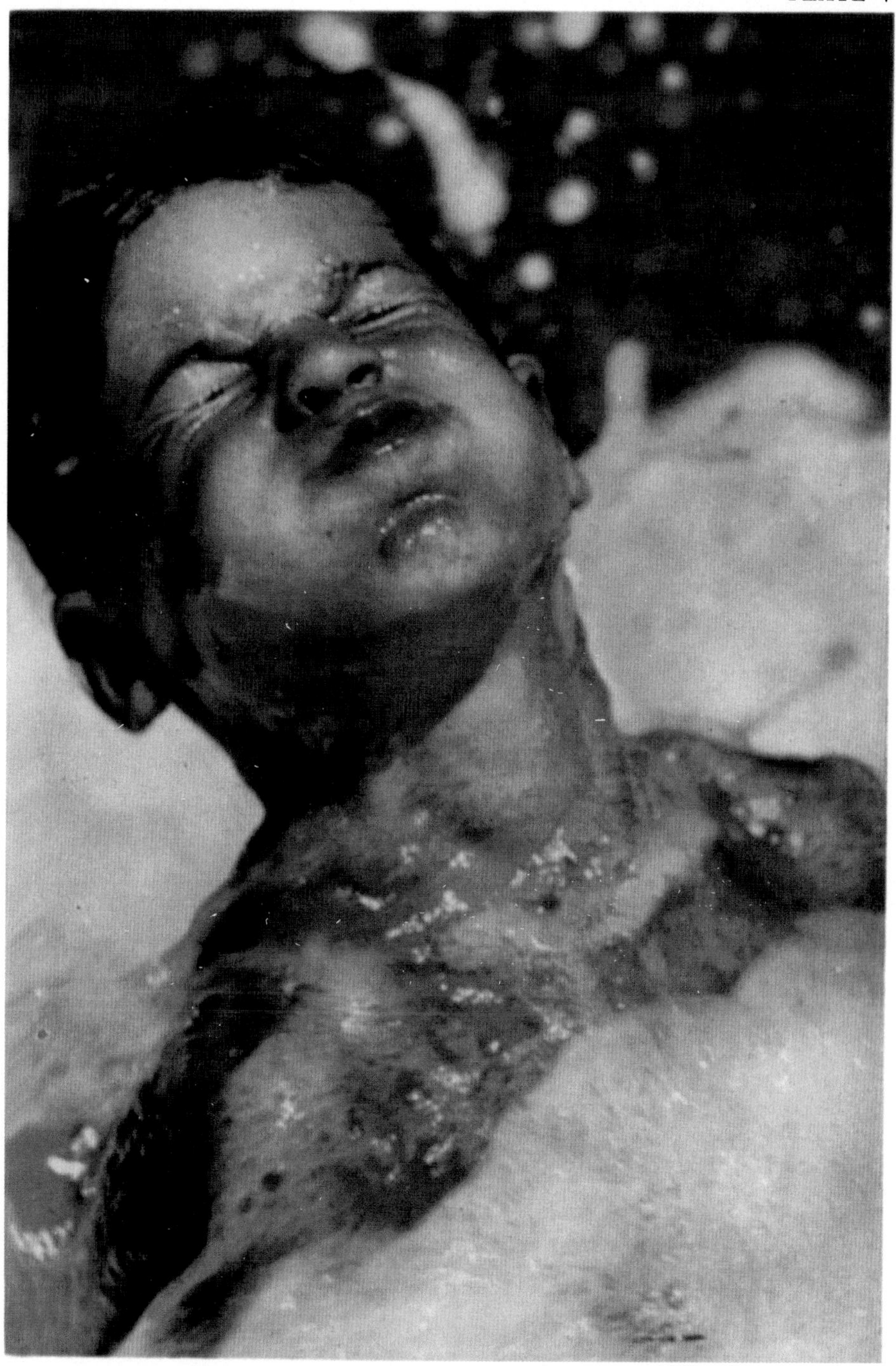

Fig. 48. Reproduction of Kodachrome picture. Taken with a miniature camera, F/4·5, $\frac{1}{100}$ second.

Cinematography in colour

Of the colour processes described, Kodachrome, Dufaycolor and Agfacolor are all suitable for cinematographic recording. The method most used commercially for entertainment films in colour is Technicolor. This is interesting photographically, since it is a quite different process. The light entering the camera is split by passage through a very thin mirror coated with a very thin film of gold. Blue-green light is transmitted, while the complementary colours are reflected on to a film which records them. The transmitted colours are recorded on two films, the first of which is blue sensitive and which transmits the green. This is registered by the second film underneath the first. From the three separate negatives, a coloured positive is made. It is thus a colour-printing process.

CHAPTER V

MAKING A PICTURE

The need for cleanliness

It must be remembered from the start that the camera is a scientific instrument. As such, it must be treated with respect and sometimes with intelligence. It should be kept in a proper case when not in use; its metal parts should be kept clean; its interior free from dust. Cameras using roll film wound on wooden spools should be cleaned carefully inside each time that a new film is put in. The smaller the negative, and the consequent need for greater enlargement in printing, the more scrupulously must care be taken to avoid dust. Films have the additional advantage over plates that while they are both packed under dust-free conditions films remain so when loaded into a clean camera; whereas putting plates into dark slides is liable to introduce dust. It is better never to take the camera into a dark room. The dark room itself is a frequent source of dirt—both dust and powdered chemicals, especially hypo. Any dark room should have one table, away from the sink, which is never used for chemicals, either liquid or solid. The hypo solution especially should be segregated because it has quite remarkable properties of creeping. It then dries up and the powder is blown about.

The towel in the dark room is often no more than a source of danger. Dirty workers wipe their hands covered with hypo or developer solution on it until it becomes a plague spot. Unless the towel is replaced by a clean one at frequent intervals, it is much better not to have one at all.

The dark room

The dark room should have its walls painted, or distempered, yellow or orange. The habit of painting walls and ceilings black is particularly silly, since the maximum amount of diffused light is required from the controlled safelight. The safelight itself can be trusted according to the directions of its makers. Incidentally, safelights are not made for the illumination of plates and papers during the whole of their development. A safelight is required to give the maximum intensity of light which is safe for the short period required for inspection. There is no point whatever in inspecting the course of development until you know that it must be nearly complete. Then maximum light is required to judge whether complete or not. The advent of panchromatic material has led to the need for carrying out dark-room operations in complete darkness. Anyone who has tried this finds that it is perfectly easy to get accustomed to doing all work in complete darkness. Maximum safety is then achieved by putting on the safelight only when development is known to be nearly complete. If exposure is known to be correct, there is no need for inspection at all. Plates are easily put into plate holders if it is remembered that they come from the makers packed in pairs, with emulsion sides facing and separated by a card slip each end.

The safelight

The object of a safelight is to provide the maximum possible light for working or for inspection of development without causing fog. A general diffuse safelight from a central hanging light may

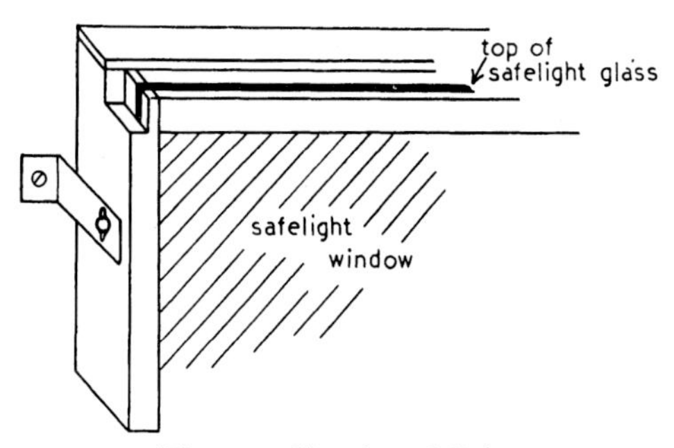

Fig. 49. Simple safelight.

be employed but, unless the dark room is used continuously for long periods, this is unnecessary. A light over the sink is sufficient.

Fig. 49 shows the design of a simple safelight box which can be made by anyone with any knowledge of woodwork. The safelight glass rests in grooves in the sides and bottom of the wooden box. As shown, it can be lifted out and replaced by other safelights as needed. The box is made light-tight by a strip of wood (not shown) which fits into the gap along the top of the box. This strip of wood has a groove cut along it, into which the top of the glass fits, and a groove across it at each end to accommodate the projecting tongue shown in the diagram. The box is fixed to any convenient support and can then be rotated by loosening the thumb screw at each side. The safelight can then be directed vertically downwards into the sink or across the dark room as desired.

It is essential that the maximum amount of light compatible with safety of the sensitive material should be used. This means that the full light should not be allowed to fall upon the sensitive material during the whole of its development but only towards the end. Mounting the light above the sink is convenient if development is carried out on a shelf or table at the side, out of the full illumination. All the reputable makers list safelights for the various types of sensitive material.[1] The box in fig. 1 holds the safelight glass in a slot so that it may be changed as required. A dark green light is essential for panchromatic emulsions. This can also be used conveniently for fast orthochromatic material. When first switched on, this light appears extremely dim, but when the eye has accommodated itself to the darkness the light is found to be quite sufficient for inspection (see p. 16). An orange light is best for bromide papers, gaslight papers and slow non-'chromatic' emulsions such as process plates. It is essential that the light source should be that recommended by the makers. Safelights are safe for a stated illumination and not safe with stronger ones. It is desirable that the number of safelight windows should be kept low, both for economy and because it is more comfortable to standardize one's work in the dark room as far as possible and always to work under the same conditions.

Washing arrangements

A fairly large shallow sink is desirable. There should be benches on each side. These are best covered with sheet lead; they can then be washed down into the sink each time that the dark room is used. It is useful to have two taps: one with sufficient clearance to fill graduated cylinders, bottles, etc., and one, for washing plates and films, which should reach down nearly to the bottom of the

[1] Cheap deep red 'safelights' are sometimes unsafe, because they transmit appreciable amounts of violet light.

sink. This one may be fitted with a gardener's 'rose'. Water should never be allowed to fall from a height into washing vessels. The wet emulsion is a delicate material. Electric light switches may be a source of danger, since the floor is usually wet and so are the hands. Dirty workers frequently shock themselves by taking their hands straight from developing or fixing solutions to the switch. In fact a really leaky switch should provide an excellent means of eliminating dirty workers on a survival of the fittest basis!

Taking a picture

Before taking any photograph it is well to ask the four questions: What? Where? When? and How? Rather obvious questions, but the beginner ignores all except perhaps the last one. In the following four sections these four points are considered and some hints given for making the best of a subject.

Before this we may consider the variables available. We have the intensity of illumination of the object. Sometimes this can be varied by altering the intensity of the light source; more usually this is fixed. Alternatively the amount of light reaching the sensitive material depends upon the aperture of the lens. In many cases this is fixed by considerations of depth of focus or by limitations of the lens. Exposure may be varied, once the above considerations have been fixed, by the range of sensitivity of emulsions available. On the other hand, the total time of exposure has often a fixed lower limit dictated by movement or change in the object. Within these limitations, it is inadvisable to use the fastest materials except where absolutely necessary, since their grain is more pronounced and the quality of the definition therefore not so good. The choice between panchromatic, orthochromatic and non-colour sensitive plates is partly a matter of personal habit and predilection and is sometimes fixed by the colouring of the object. In general it is highly desirable to work always with the same material so far as possible. The type which covers all but exceptional cases (such as extreme ultra-violet and infra-red) is the ordinary commercial fast panchromatic (Kodak Panatomic film and Ilford Special Rapid Panchromatic plates). These require handling in the dark room in complete darkness, but this is very easily learned. It is desirable to have a standard technique in the dark room and, therefore, to stick to one type of emulsion. There are very few objects to be recorded in scientific work which cannot be done on the two standard materials—panchromatic and, for drawings, a process emulsion.

Object matter and direction of view

Before taking any photograph it is necessary to decide what is to be taken and what is to be left out. This seems to be a fatuous remark, but any experienced photographer knows that more photographs fail through ignoring this elementary precaution than from any other cause. It may be objected that unwanted objects can be cut out afterwards and the object required enlarged so that nothing is lost. This, however, is not true, since getting unnecessarily far from the main object to include things around it frequently makes it impossible to get the perspective which gives the most pleasing result. This is particularly important where apparatus is being recorded. The rigid elimination of all objects except that of basic interest is essential. If one has a piece of apparatus of novel design which is worth recording, why spoil the effect by reducing its size and distracting attention from it by inclusion of the various accessory measuring instruments, etc. around it with whose appearance everyone is already familiar? The best results are always obtained by including in the picture as little as possible. This also applies to the background, which frequently introduces objects not required, such as benches, conduits, etc. Fig. 50 (*a*) shows the bad effect of an obtrusive background. It cannot be emphasized too strongly how easy it is to overlook such things when the picture is being taken. Even when focusing is done on a ground-glass screen, attention is so fixed on the principal object that the rest is overlooked. With apparatus that cannot be moved, the background can be eliminated to some extent by sheets or similar screens. Fig. 50 (*a*) shows the use of such a screen and white paper on the floor.

The difficulty is often encountered in photographing apparatus that the most prominent object is the housing of the apparatus. This is always accentuated by a normal direction of view. The picture contains the rectangular lines of a box or whatnot which distract attention from the principal object by leading out of the picture on its four sides. This is easily evaded by choosing a viewpoint which is not normal to the apparatus—on one side and possibly from above or below. More detail of the apparatus is

Fig. 50 (*a*).

usually shown by one of these viewpoints. Figs. 50 (*a*) and (*b*) show two pictures of the same piece of apparatus which illustrate a number of these points. Fig. 50 (*a*) shows the complete set-up. The essentials are lost and the picture is of interest only as a record for personal use and of little value for publication. Fig. 50 (*b*), on the other hand, shows some of the essentials and eliminates entirely, by its worm's eye view, the objectionable obtrusion of the supporting cage of the apparatus. Note also in fig. 50 (*a*) the presence of various distracting odds and ends around the apparatus. As many as possible were cut out by the screen but some remain. Remember always that the camera includes everything in the solid angle of its vision. The human eye here again misleads persistently and deceives even the most experienced photographers. Objects on the plate are ignored because the attention is concentrated on the principal object. This is automatic in the case of photographing a specific object. It is unbelievable what objects may be ignored even when the image is viewed on a ground-glass screen simply because the attention is being given to focusing the principal object and others are ignored. If you don't believe this, examine the prints of any amateur and note how many (in which there is one principal object, such as in portraits) there are which show peculiarities due to unwanted objects in obtrusive positions, e.g. trees growing out of the sitter's head or the sitter's face cut in half by the top of a wall in the immediate background.

The second question Where? is closely bound up with the first. It resolves itself into the two possibilities of moving the object or the camera to a suitable place. Usually, however, the object is fixed so that choice of a suitable viewpoint is all-important. Where the object consists of predominantly rectangular lines, a normal viewpoint produces a picture in which the main lines cut the picture into sections which have no continuity or coherence. In the case of buildings, the main horizontal line is apt to cut the picture clean in half. A flat horizon does the same thing unless objects in the foreground are brought in to save the composition. In the case of buildings, the unpleasant effect may be eliminated by taking the photograph from one side, so that the principal lines form a wedge-shaped pattern which is much more restful to the eye. Incidentally, this viewpoint gives a much better three-dimensional impression than does the normal one.

Fig. 50 (*b*).

Direction of illumination

The question When? is usually settled by expediency. In outdoor work it is very important to wait until the sun is in the right direction for the picture. Having chosen the viewpoint which shows the shape of the object best, it may be found that the sun is shining from the correct angle for a very short time only each day. Always the hours from 11 to 3 should be avoided as far as possible, since results are poor owing to the overhead position of the sun and consequent shortness of, or absence of, shadows. Whatever the position of the sun, it should never be directly behind the camera, since flat lighting results. In general an angle of 75° to 150° between the direction of the sun's rays and that of the camera is good. Bigger angles may be used profitably sometimes, provided the lens is protected by a hood, and exposure is increased for the absence of light on the camera side of the objects in the picture. Fig. 39 shows an example of photographing against the light (*contrajour lighting*) to give 'guts' to a dull subject.

Composition

When the eye looks at a print, it is doing something unnatural. It is fixed in focus on one plane, whereas in examining the original it is wandering from far to near and back again continuously. It cannot do this on the print but it still tends to wander in the plane of the picture. This explains some of the simple rules of composition. A principal object should never be placed in the centre of the picture but to one side. The eye then tends to wander from the centre back to the principal object, whereas if this is in the centre, the eye has the choice of four sides of the picture to which to wander. Square pictures are the worst, since there the eye has four equidistant sides to which to wander, whereas in the rectangular picture the two nearer ones are preferred. For the same reason marked rectangular lines are always bad, for they attract attention from whatever may be the real principal object, and lead out of the picture. Pictures cut in half by sky line or other division are bad since the eye is led out each end. The ideal is such an arrangement of objects and main lines as leads the eye round and round about the principal object. Even where the picture consists of a single object against a blank or indeterminate background, and not of a composition of objects and lines, the parts of the object must obey the same rules, e.g. hands and arms in a portrait must lead back to the face or else be cut out. In brief, the eye will not fix itself upon the main object of the picture unless it is forced to; and, conversely, if there is anything in the picture to lead the eye astray, it will do so and take attention from the principal object.

Fig. 50 (*a*) shows curiously well the antithetic tendency of the objective compositional effect and the subjective aesthetic one. The picture is definitely pleasing, although bad as a record of the essential parts of the apparatus. This is because the rectilinear lines, especially the verticals of the cage, give an impression of strength and stability which is, of course, ideal in a set-up of that sort. Although the composition of lines is bad, it is saved from fussiness by the lighting which accentuates the principal object and uses the dull brick pillars to lead the eye to the more brightly lit principal object.

Viewpoint and perspective

Fig. 51 shows the so-called false perspective (see also p. 32). When the picture is viewed from the correct distance—the distance equal to its diagonal—the perspective is correct. If, now, it is viewed from a distance considerably greater, the perspective becomes more and more false as the distance increases. It is the viewing distance which makes perspective 'true' or 'false'. This picture also raises another interesting complication. The proper viewing distance and therefore correct perspective give a picture which corresponds with a close viewing position of the original object. In some cases, there is a sub-conscious tendency to choose one's viewing distance so that the picture has the perspective of common experience of that particular object. In this case, the customary distance of viewing is considerably greater than the distance of this view. The viewer is therefore at cross purposes. If he takes the position dictated by his own experience he will get false perspective. If, however, he takes the proper position (equal to the diagonal of the picture) he gets correct perspective.

The question How? is answered more or less diffusely throughout this book, but a few special points may be mentioned here. In the first place, never forget the uncanny capacity of the eye for deceiving you. First, view the subject through one eye only to eliminate the stereoscopic effects of binocular vision. Secondly, view the subject with the eyes partially closed so that the main objects only are seen somewhat fuzzily. Their arrangement and ability to form a pictorial arrangement are now more easily judged. Next, decide how much of the effectiveness of the scene is due to

Fig. 51. Perspective correct when viewed from distance equal to diagonal; incorrect when viewed from greater distance.

colour contrasts. Some idea of this can be gained by viewing through a yellow filter or, if non-colour sensitive material is being used, through a blue one. Elimination of colour contrast makes it essential to force principal objects to stand out by their placing and lighting alone. The boldness of the principal object can frequently be increased by use of a large stop giving small depth of focus which leaves the unwanted background fuzzy and not distracting. This method can be used most effectively when the principal object is close to the camera. When farther away, aerial perspective can be used to produce the same effect.

Aerial perspective

Fig. 52 shows the basis of aerial perspective. If one considers receding planes it is evident that the area from which light is concentrated on to the plate increases as the square of the distance of the plane. On the other hand, the amount of light from any given plane falls off as the square of the distance from the camera by the ordinary inverse square law. This is easily seen if the illumination of any plane is regarded as a distribution of point sources. The smaller diagram shows the limitation of the amount of light received by the lens as distance increases. Since these two factors cancel out, it is evident that the brightness of equally illuminated receding planes should not vary with their distance (see p. 41). This is obviously contrary to experience. The reason is that scattering of light occurs in the atmosphere between the plane and the camera. Extra light reaches the camera because the light striking the dust and water particles from directions outside the angle of view of the lens is scattered in all directions according to the Rayleigh equation (p. 50) so that some of this light reaches the camera. The amount of extra light reaching the camera is obviously, under any given set of conditions, proportional to the distance between the camera and the plane under consideration. Owing to this admixture of light from outside the angle of view, colours become degraded according to their distance. The amount of scattering depends upon the amount of suspended matter in the air and on the direction of the sunlight falling upon it according to the Rayleigh equation. It is also clear that the shorter wave-lengths, violet and blue, will be preferentially scattered.

Aerial perspective is very valuable to the pictorial worker. It is obviously most marked when the principal object is close to the camera and the background remote. The extent of aerial perspective due to scattering is affected by filters, since the maximum scattering is in the shorter wave-lengths, violet and blue (see p. 52). The concentration of light by distant planes may be a nuisance to the record photographer, who finds unwanted backgrounds un-

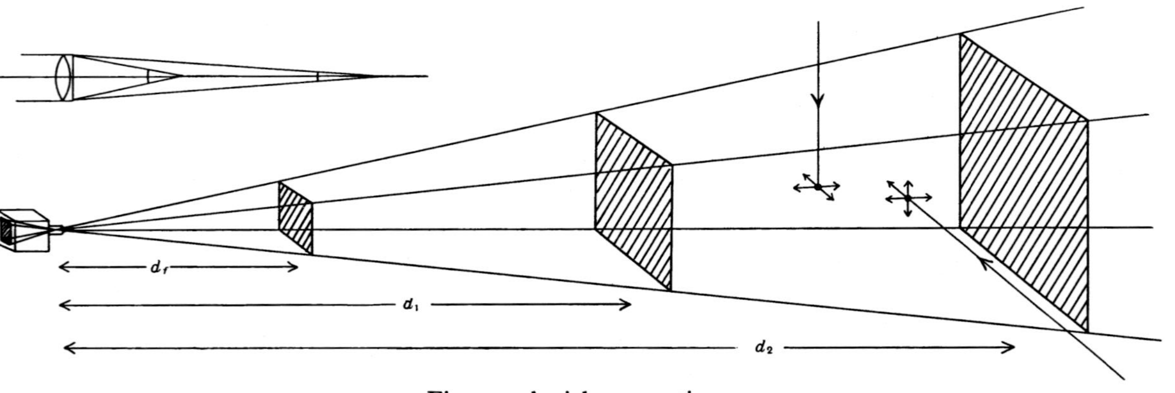

Fig. 52. Aerial perspective.

Fig. 53. The Kremlin, Moscow. Showing ordinary and aerial perspective.

expectedly bright. For the same reason, light objects cannot be photographed against a light background unless the latter is differentially darkened by suitable filter.

Fig. 53 shows an example of ordinary perspective in the foreground with aerial perspective in the distance. It was taken about

Fig. 54 (*a*).

sunset with some mist over the river, so that great distance was not needed for the effect. Panchromatic material with fully correcting filter was used, so that the picture gives the same amount of aerial perspective as was seen by the eye. Note how badly fig. 54 (*a*) requires light-scattering as in 54 (*b*) to dull the obtrusive background.

Lighting

Fig. 54 shows an instructive example of the effect of changed lighting. (*a*) was taken in full sunlight with the clear sun high in the sky. (*b*) was taken from the same viewpoint in the evening when the sun's disc was covered and all the light was diffused. It

Fig. 54 (*b*). The Cathedral, Antwerp, taken from the Boerentoren.

was fully exposed and the print looks flat because the subject was flat. Contrast of light and shade requires directed lighting and the full effects can be obtained only when the sun's disc is uncovered. When a single object is photographed, perspective does not help much to show its three-dimensional shape unless it is very large. This can be done only by suitable lighting. If the light comes from behind the camera, the object will look flat. The intensity of light must be greater from one side. The difference of intensity varies with the type of subject. For portraits, the difference must only be slight. Portraits taken in full direct sunlight are always bad because of excessive contrast. If there is not a cloud over the sun, diffuse light reflected from a convenient wall is good. When, as is usually the case, the sun is high in the sky, portraits in its direct light always show a skull-like effect because no light reaches the eye sockets.

Artificial lighting

Portraits in artificial light require the same care to get inequality of lighting on the two sides of the face but without excessive contrasts and without sharp discontinuities of lighting. The safest general method is to provide the general light uniformly over the face from one lamp with a second lamp closer to the subject and placed unsymmetrically with respect to the direction of the camera. It must be to one side and may also be above or below the centre, according to the needs of the subject and the effect required. Hardness is avoided by covering the light source with a diffusing screen which converts the point source into one of finite size.

Colour filters and their uses

Filters can be made in three forms:

1. Parallel-sided glass cell containing solution.
2. Dyed gelatine sheet (enclosed for protection between glass plates).
3. Coloured glass.

Of these the first is convenient for laboratory work on account of the ease with which solutions can be changed or their concentration altered. As already pointed out (p. 46) the thickness of such a cell has a definite effect upon the optical path of the light and can upset focusing. Where visual focusing is employed through the filter, this error is eliminated. For most photography, gelatine sheet is used. This can be bought in squares or circles and used without protection, but it becomes scratched quickly. Mounted between glass, it is quite permanent if not ill-treated. Coloured glass filters have the advantage of permanence but the drawback that the range of colours is limited, whereas the choice of dyes which can be used as gelatine sheet is very large. Metal filter holders to slip over the lens are supplied by the manufacturers.

For a full list of the filters available reference should be made to the Wratten and Ilford lists. Those used most commonly are

Fig. 55. White flowers against blue sky through deep yellow filter.

pale yellow or yellow-green to give an approximately correct tone rendering, on panchromatic emulsion, of colours in white light. The paler yellow ones require 1·5 to 2 times increase of exposure on panchromatic film, and do not give full correction, but on this and on orthochromatic material they are useful in landscape scenes for cutting down the intensity of the blue sky light and rendering white clouds. For this purpose, they are sometimes made as 'sky filters', in which only part of the disc is coloured, so that the exposure for the ground part of the picture is not cut down. For special effects, such as exaggerating clouds in a blue sky, red or orange filters are used. They lead, however, to aesthetically unpleasing results in many cases by overdoing the correction. In certain places, such as mountains, there may be an excess of ultra-violet light, which should be eliminated by an aesculin filter. This, however, has the drawback of fluorescing and emitting blue light, which causes some veil on the negative. The Pola screen can be used to cut down the intensity of the blue sky light provided it is used at the proper angle to the direction of the sun's rays. It has the advantage of not cutting down the rest of the illumination proportionately. It is the only type of sky filter that can be used with colour processes.

Fig. 55 shows the use of a yellow (3 times on panchromatic) filter to show white flowers against a blue sky. Some workers use a filter to give correct tone reproduction habitually and without consideration of the objects to be photographed. This is often bad, since contrasting colours may be rendered in the same tone, e.g. reds and greens. Again, where it is desirable to make use of aerial perspective, no filter should be used, or a blue-violet one, since, as already explained, haze is caused predominantly by scattering of violet and blue light. Conversely, a red filter of about 8 times exposure is useful for cutting out haze without the unpleasant effects associated with infra-red. Fig. 12, Chapter 1, shows a view from the top of the Eiffel Tower taken in summer when there was a considerable amount of haze. Penetration by the photograph is considerably greater than visual; visually the Sacré-Cœur Church was seen only hazily and nothing of the distant horizon. This may be compared with the infra-red picture, fig. 10, Chapter 1, which shows the characteristic 'snow' effect of green foliage (see p. 24). The trees in fig. 12 are much more naturally rendered. A 3 times yellow filter on panchromatic emulsion gives about the same amount of penetration as the human eye, perhaps a little more.

Another filter used in all types of work is the blue one used to correct the yellow colour of artificial light. These are often called half-watt filters but must be used with care, since different varieties

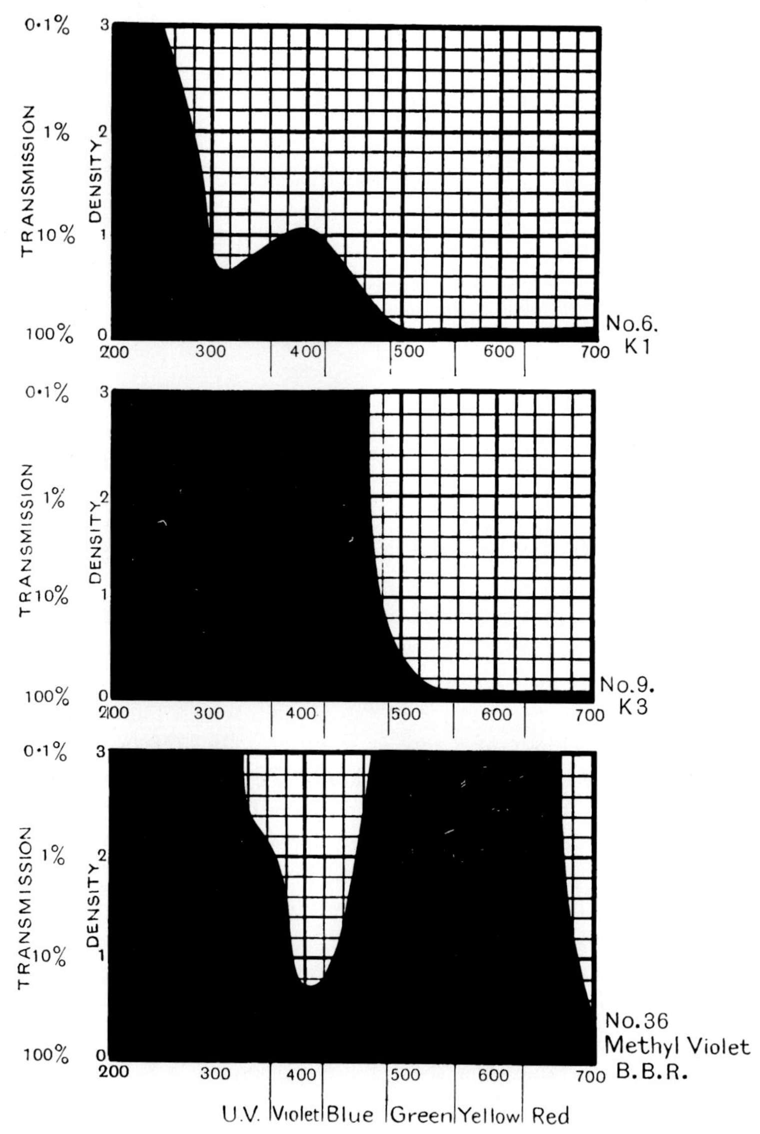

Fig. 56. Light transmission and absorption by filters.

and even different sizes of the same make of lamp vary considerably in running temperature and, therefore, in the amount of correction required. These blue filters are especially important for colour photography by artificial light. The following table gives the approximate running temperatures of a number of common light sources:

Sunlight	5400 K.
White flame carbon arc	5000
Ordinary carbon arc	4000
Photoflash	3600
Photoflood	3480
Half-watt lamp Tungsten, gas-filled	3000–2700 (according to size)
Acetylene	2360
Carbon lamp	2080
Candle	1930

Fig. 13, Chapter 1, shows the energy distribution of colours in white sunlight, blue sky light and half-watt electric light.

In scientific work, correction by filters is much less common than their use for deliberate distortion to increase contrast or for selection of a small part of the spectrum.[1]

Fig. 56 shows absorption of light by three Wratten filters. The K_1 and K_3 are pale and deep yellow ones used for ordinary landscape correction on panchromatic emulsions. The methyl violet one is included as an example of the caution which must be used in judging the colour of a dye or filter visually. Often, there is transmission in an unsuspected region of the spectrum in addition to the main colour transmission band. Here there is an appreciable amount of red as well as the spectrum violet. Fig. 57 shows the absorption curves of some filters used in scientific work. The top set is of filters for correct tone rendering. The tri-colour filters in the middle set are those used in three-colour printing to separate the whole spectrum into three parts, and the bottom set is for use in micrography or other subjects where a narrow transmission band is required. Special filters are also made for segregating monochromatic radiations from a line discharge, e.g. a mercury vapour lamp or a cadmium arc.

[1] Lists of filters manufactured are issued by Messrs Ilford (*Panchromatism*, 6*d*.) and Messrs Kodak (*Wratten Light Filters*).

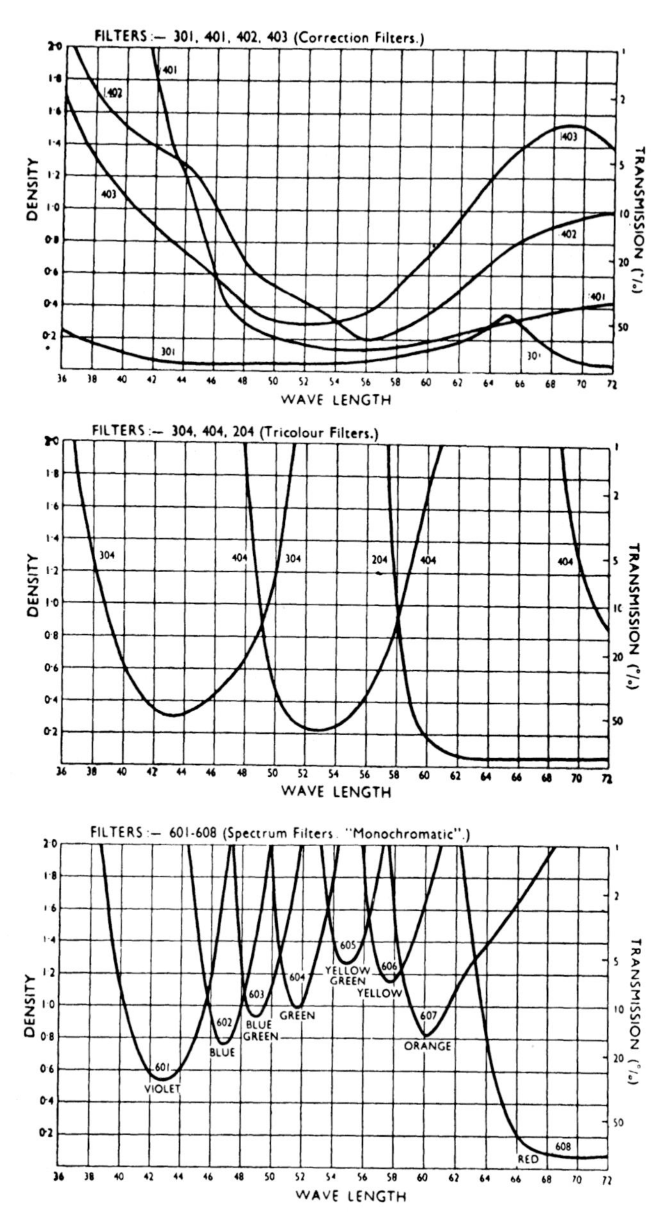

Fig. 57. Light transmission by filters.

Speed of emulsions and exposure

The larger firms of manufacturers provide ample working details with their plates, films and papers and several of them sell books of instructions suitable for those with little knowledge of photography. One word of warning, however, must be given. Speeds of emulsions and development times are not to be trusted too implicitly. The situation concerning these has been in a state of confusion for some years now as an unfortunate relic of the period in the early nineteen-twenties when the speed of emulsions suddenly began to increase rapidly. As soon as each manufacturer found a faster emulsion, he put it on the market, usually with a fancy name, so that quite soon the available names indicating high speed were used up long before present speeds were reached. Plates still exist on the market with names such as 'extra rapid' which, by present standards, are classed as medium slow. Difficulties arose over the difference of form of the characteristic curves of the newer materials and new methods of rating were devised as improvements on the old H. and D. system. To make confusion worse, English ratings on the H. and D. method are slower in comparison with the H. and D. ratings used by continental manufacturers by a factor of about 3 times (see p. 13).

An absurd practice which has arisen is the rating of panchromatic and orthochromatic emulsions with two speeds—the normal H. and D. and another 'H. and D. speed to half-watt lighting'. The method is to compare speeds of the emulsion to half-watt illumination and to compare the result with a non-colour-sensitive plate of some other H. and D. rating. The colour-sensitive plate is then given its half-watt speed by multiplying the low speed by the ratio of sensitivities to the coloured light. No further comment is necessary!

A second very objectionable cause of poor results is the excessive time of development recommended by manufacturers. This seems to be partly a relic of the speed race and partly because over-development does not matter for contact prints. In fact it helps to give an appearance of extra sharpness to the prints and the general public have become educated into accepting 'soot and whitewash' prints to such an extent that they do not ask for anything else. When, however, enlargements are required, the effects of over-development are very serious. Full tone value is lost and, in particular, sky and high-light detail is all but lost by blocking-up of the corresponding parts of the negatives. With the formula given on p. 168 results correct for most plates and films for printing on medium-grade bromide paper will be obtained at 65° F. in about 2 to 2½ minutes. The correct times of development vary with the make of film and its speed. All workers should determine for themselves the time required to give a negative that will print on a medium-grade bromide paper. This can be done conveniently by desensitizing the first few films or plates and judging the end of development by a bright light. The end of development is judged by the beginning of veil over the lighter parts of the negative, assuming, of course, that it is fully exposed. It should be noted that all materials appear much darker in the dim light of the dark room than they do in full daylight. The old workers, who worked largely by means of 'tips', developed their prints until they thought they had reached the right density and then gave them another 30 seconds. The multiplicity of types of materials and differences of working methods render more accurate methods necessary to-day; time and temperature should always be used for development after making one's own standards (see p. 110).

Actinometers and exposure meters

Under this heading are included three types of meter: actinometers proper, photo-electric cells and extinction photometers. The actinometer, as its name suggests, measures the actinic value of the light. The sensitive material is a piece of slow bromide paper which is exposed to the light until it has darkened to a standard tint. From this time the correct exposure for an emulsion of any known speed and for any required stop number can be calculated. In weak lights, such as interiors, the time required to darken to the standard tint may be very long, but this can be compensated for by adjusting the meter to the stop number for which time of darkening is equal to time of exposure required and then carrying out test and exposure concurrently. These meters have gone out of fashion, partly because the sensitive paper is not panchromatic, which makes them unsuitable for use with panchromatic films, and perhaps more because they are most useful in bright light, whereas exposure meters are needed most, especially with modern large aperture lenses and fast films, in feeble or artificial light. Their accuracy is limited by the observer's ability to judge equivalence of the test strip and the standard. The photo-electric cell consists of a small galvanometer which records the current emitted by the cell, when illuminated. By means of scales and tables the correct exposure is read off from the deflection. The spectral sensitivity of the photo-electric cell differs in shape somewhat from that of the typical panchromatic emulsion. Allowance should be made for this towards sunset (fig. 11, Chapter 1).

The extinction exposure meter consists of a grey filter of continuously varying depth through which the subject is viewed directly, and the instrument is adjusted until the subject can no longer be seen sharply. A scale is attached to show the depth of the filter and, from it, the necessary exposure calculated. These meters are, of course, unsatisfactory in weak light and have the general objection that the reading depends upon the subjective judgment of the observer. They have the great advantage, however, that observations can be made on any part of a subject.

All exposure meters are liable to give grossly incorrect values unless used with intelligence. The angle of view is not usually sufficiently restricted, so that too much light enters them and an unduly short exposure is indicated.[1] Care must therefore be taken to shade the meter from sky light. In any case, care must be taken to obtain a reading from the actual subject to be photographed and nothing else. When this is done meters are extremely useful. Correction must be made, of course, for filters.

A useful and accurate method of using an exposure meter, especially for interior work, is to place a sheet of dull white card in the plane of the object to be photographed and to direct the exposure meter at this at a distance sufficiently close for the card to cover the whole angle of view of the meter. The value indicated is taken as the light intensity of the subject. If a meter is used directly in an interior, it will usually include some light entering by windows. The value of the exposure indicated will then be too low for the general illumination of the room. A little consideration shows that a white card should be used whatever may be the colour of the objects to be photographed.

[1] The latest model Contax camera has a built-in exposure meter which is provided with vanes which restrict the light falling upon it to the correct angle. It also has a cover which, when lifted up, forms a sky shield.

Exposure of moving objects

The maximum exposure of a moving object which will give a sharp image is obviously determined by the permissible diameter of the circle of confusion. The movement of the image during the exposure must not be greater than this. The formula

$$\frac{s_o}{s_i}=\frac{o-f}{f}$$

gives the ratio of size of object, s_o, to size of image, s_i, in terms of object distance, o, and focal length of the lens, f. For an object moving at v ft. per sec., at right-angles to the line of sight, we may substitute for s, Ev, where E is the exposure. E_{max} (when the image movement is equal to the diameter of the circle of confusion) is then given by the equation

$$\frac{Ev}{\delta}=\frac{o-f}{f},$$

where δ is the diameter of the circle of confusion, or

$$E_{max}=\frac{\delta\,(o-f)}{fv}.$$

Longer exposures may be used when the motion is at an angle to the line of sight, because we substitute $v \sin\theta$ for v, where θ is the angle between direction of motion and line of sight. This method of calculating maximum exposure may break down when the object is close to the camera, since the outlines of the image are moving, owing to alteration of size during the exposure, in addition to its real movement.

The table below shows the approximate maximum exposures permissible for subjects moving in a direction at 45° to the angle of view at a distance of 100 focal lengths of the lens used.

Subject	v	E_{max}
Walker	4 m.p.h.	1/75 sec.
Runner	12·5	1/300
Horse, trotting	10	1/250
Horse, galloping	20	1/500
Waves	—	*c.* 1/500
Car	35	1/750
Car, racing	60	1/1250
Express train	60	1/1250
Aeroplane	120	*c.* 1/3000

Fig. 58. Russian horseman.

The exposure will be the same if the camera is moving at these speeds and stationary objects are being photographed at the same distances.

The velocity of parts of an object may be very much greater than that of the object. The feet of walkers provide a common example. Fig. 58 shows the same effect in a horse. In many moving objects,

Fig. 59. Position of zero velocity.

the motion is discontinuous and periodically becomes very small or vanishes. Fig. 59 gives an example; at the top of each swing, the motion ceases for a short time.

If it is necessary that the principal object only should be sharp, the camera may be swung round to follow the moving object, and then a much longer exposure will give a sharp image provided there are no parts moving with independent velocities in other directions. When the camera is swung, the background is fuzzy, usually to the advantage of the picture, since attention is concentrated on the main object and an impression of speed conveyed without blurring the moving object.

Many photographers, especially those using cameras equipped with some form of visual focusing such as reflexes and cameras with coupled range-finders, have false ideas on the subject of focusing moving objects. It is obvious, indeed, that so long as focusing takes a finite time, the object will have moved during that time. The only method which is satisfactory, even for quite slowly moving subjects such as pedestrians, is to focus the camera on some convenient object at the desired spot and to make the exposure when the moving object passes it. The only further point to remember is to avoid jabbing the release down hard and moving the whole camera in anxiety to catch the object just on the spot. There is a small but definite time lag between deciding to press the release and actually doing it. With very rapidly moving objects, this may be serious.

Correct tone reproduction

Papers appear at frequent intervals on the correct reproduction of tone values and, from what has been said already, it is clear that direct reproduction requires suitable fitting of the proper parts of the characteristic curves of negative and printing paper. The academic question of 100 per cent accuracy of reproduction is really a red herring in the majority of photographic work.

The range of contrast of illumination in a sunlit landscape may be as much as 60 to 1. In heavy mist it is not more than 2 to 1. Fig. 54 shows the marked difference of range of contrast between direct illumination by sunlight and diffuse lighting. Heavy cumulus clouds on a bright day still further increase the range of contrast. The reproduction of such effects in a positive print is limited by the range of contrast possible on the paper. Lamp-black on white paper gives a range of contrast of only 20 to 1. This is because the black is not truly black and the white does not reflect all the incident light. White paper reflects from 60 to 80 per cent of the incident light. A glossy print can give a 50 to 1 range of contrast but a matt print is limited to 20 to 1. This is because the black in the matt print is very far from a true black, i.e. one which absorbs all incident light. With positive transparencies, the range of contrast may be very much increased, since the amount of light passing through the high-lights may be increased at will by increasing the power of the light source.

Examples have already been given of cases where gross misrepresentation of tone values is required to show up details of structure. Even in straight record photography, it is seldom required to produce an accurate tone rendering since, in the first

Fig. 60. Scene of great lighting contrast. Lower tones sacrificed to give greater range of contrast to the medium and light ones.

place, it is impossible by any means whatever to produce a paper print which shows a range of intensity as large as that commonly occurring in sunlit subjects. In fact, it is much more usual that the photographer needs to decide, after making a trial print from his negative, what tones he is going to sacrifice or, more accurately, which end of his scale he is going to condense. Fig. 60 shows well the value of sacrificing the dark tones in a subject with very great contrasts. The high and medium tones are extended to give

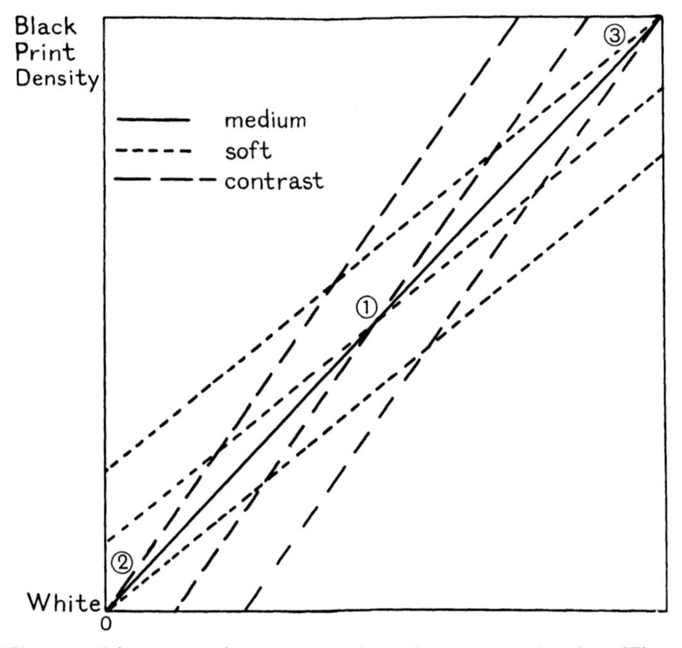

Fig. 61. Diagrammatic representation of tone reproduction. The numbers refer to the lines which intersect at them.

them 'body' and the dark tones are all crowded into black. One may compare the characteristic curves of soft, medium and contrasty papers by making them intersect at points chosen anywhere in the tone value scale according to whether it is desired to get maximum tone range in light, dark or medium parts of the negative, and to see also the effect of extending or condensing the extreme high-lights and shadows (fig. 61). In this connection we may note that it is commonly suggested that combined use of panchromatic emulsion and suitable filters will give correct reproduction of sky and landscape. This is impossible. A filter can reduce the excess of blue light coming from the sky, but it cannot alter the intensity ratio between, say, a bright white cumulus cloud and deep shadows in the landscape. In printing we must compromise. If the clouds are shown reasonably correctly, then the ground details will be too dark for accurate reproduction although aesthetically the result may be perfectly satisfactory. Alternatively, if it is desired to show a landscape in a very high key corresponding to its appearance on a summer day, it will be impossible to render clouds at their full value against the blue sky. In fact, we have the choices shown diagrammatically in fig. 61. This shows the idea of matching paper to negative. Straight-line relation between image density and exposure is assumed for simplicity and it is also assumed that we are printing from a negative of gradation fitted to the medium contrast paper. This is denoted by the full line. Prints made from this negative on soft and contrasty papers will alter the tone distribution in the print from that in the negative. From either of these papers we can get any number of parallel lines in this diagram according to which part of the tone scale is sacrificed. In the diagram, three cases are shown. In curves 1, exposure is given such that the middle tone is the same for the three prints. Then the soft paper print will have no clear whites and no full blacks while the contrasty paper will have some of the light tones crowded into white and some of the darker shadows, as shown, full black. In curves 2, exposure is made to give correct clear whites and the shadows left to look after themselves. The result is that the soft paper contains no real blacks at all, only greys, while the contrasty paper crowds still more shadow detail into full black. The lines marked 3 show the reverse of this last case; the shadows being used to determine the exposure and the high-lights left to look after themselves. With soft paper, we get no clear whites, only dingy greys as highest tone, and with the contrasty all high-light detail is lost as it is shown as clear white. Of the prints shown in fig. 62 (*a*) is case 2 for soft paper; (*b*) is nearly case 1 for soft paper; (*c*) is case 1 for contrasty paper and (*d*) is the medium paper over-exposed.

We see, therefore, that correct tone reproduction of a sort is theoretically easy, but this is not really correct reproduction, since

Fig. 62. Incorrect tone reproduction by use of wrong papers and exposures. All from the same negative. Fig. (*c*) is the correct tone reproduction.

the original brightness distribution is reproduced on some arbitrary scale—bearing a simple relation to the original it is true, but nevertheless different. The subject is further complicated, once more by the human eye, because the relation between objective and subjective brightness is not linear. The reader who desires further information is referred to a very interesting paper on the subject by R. G. Hopkinson (*Biochem. J.* Sept. 1937, p. 542).

Routine directions for making an exposure

Before loading, make sure that the interior of the camera is free from all dust. See that the rollers in a roll-film camera are smooth and clean. Do not ever put a roll film or film-pack into the camera in full sunlight. If in the open, the shadow of a tree or bush or one's own shadow is sufficient. When the free end of the roll-film paper is tucked into the take-up spool, give the handle a couple of turns, shut the camera and wind steadily until the numbers are seen through the window. The winding should not be done jerkily. If panchromatic film is being used, the red window must be protected from direct sunlight. Fast plates taken out of their original wrappings deteriorate even in darkness after a few weeks.

Before using the camera, make sure that the lens is clean and free from dust. Then decide which of the adjustments are fixed by the requirements of the subject—depth of focus, rate of motion, brightness of light and whether colour filters are needed or not. Then measure the light intensity by a photo-electric cell or, if you have not got one, use one of the numerous tables available which give the light intensity at any required time of day and under various weather conditions.[1] Make sure, as described more fully previously, that the object that you want to record is not confused by other objects in the angle of view of the camera, because these will all be included in the picture. A suitable view-point being chosen, the exposure may be made. Take a deep breath and press steadily on the release. That does not mean lean on the camera. Use one finger or thumb with a counter pressure from the rest of the hand. Exposures up to a second may be made with a

[1] For recording subjects indoors, I find that when the image on a ground-glass screen is just not clearly defined, about 10 seconds is required at *f* 4·5 on a plate of speed H. and D. 400 (English). By stopping down brighter subjects to this point, their exposures can be calculated for any stop required.

hand camera but without any certainty of holding it steady. When making exposures of the order of 1 or 2 seconds, releasing the shutter or removing a lens cap is liable to set the camera into motion which will last during the exposure and spoil definition. This is easily avoided by holding something in front of the lens but not quite touching it, making the release, waiting for any motion to stop and then making the exposure by the independent object. This may be a sheet of card or, out of doors, one's hat. If the photographer is out of breath, it is most unlikely that he will be able to hold the camera steady for longer than $\frac{1}{50}$th second. After making the exposure, wind on the film immediately. If a plate camera is being used, there is the additional item to remember of removing the shutter from the dark slide. It is very easy to forget this, especially when one is not used to the camera. Plate holders are usually numbered consecutively and should always be used in numerical order.

These are all matters of routine, and unless they are done almost automatically, the photographer has little chance of attending to the more difficult finer points of getting really first-class results. The manufacturers supply booklets of instructions for their own particular products and these should be studied carefully by the inexperienced worker before trying to use the camera. It is very useful to familiarize oneself with the camera movements before putting in a film or plates.

CHAPTER VI

DEVELOPING AND PRINTING

Making up and keeping solutions

Solutions need not be made up in distilled water; tap water, unless especially hard, is good enough. Care, however, should be taken that it does not contain dissolved air, since this is liable to form air bubbles on negatives. Solutions may be kept conveniently in Winchesters with rubber bungs. It is not safe to trust ground-glass stoppers for developer solutions. In the Appendix, p. 167, several developing and fixing solutions are given. In making up solutions the makers' directions should be followed. Intelligence, however, may be used in this matter. The developing agent should be weighed with some care and the bromide with great accuracy. Carbonate and sulphite need little care. In making up small or medium amounts of developer, the weight of bromide is usually rather small for accurate weighing, and it is much more convenient to make up separately a 10 per cent solution and to add to the developer the volume containing the desired weight. As regards order of dissolving chemicals, reducing agent should be added first—if metol and hydroquinone are both present they should be added in this order. Then the sulphite, followed by the carbonate. The bromide can be added at any stage. A little metabisulphite may be dissolved in the water first of all to remove dissolved oxygen. Warnings are often given against warming solutions to facilitate solution of the chemicals. This warning is quite unnecessary in the earlier stages. If water heated to about 45° C. is employed, addition of the sulphite reduces the temperature and by the time that most of the carbonate has dissolved, and the developer is activated, the solution has fallen to room temperature. By starting with warm water, the dangers due to dissolved air are eliminated. Most manufacturers now control the size of the crystals of carbonate and sulphite and supply them as 'pea crystals'. These are most convenient for ease of solution. Sometimes advice is given to use the monohydrated sodium carbonate powder. In general, however, anhydrous powders are not good for ease of solution, since they tend to form lumps in water.

Hypo should always be dissolved in warm water, since solution is accompanied by considerable cooling. Metabisulphite should not be added to hot water as it is then decomposed.

Desensitizing

Panchromatic emulsions may be desensitized by certain dyes, of which pinacryptol green, phenosafranin,[1] and basic scarlet are the best known.[2] Development may be carried out in bright yellow light after bathing the negative in the dye solution for 1 minute. Fast panchromatic emulsions can be developed in the light of a candle flame after desensitizing. Alternatively the desensitizing dye may be added to the developer, the earlier stages of development carried out in complete darkness and the completion of development judged in the bright light. The second method has the disadvantage that a new lot of dye has to be used for each development. By the first method a desensitizing solution may be used over and over again. Desensitizing, however, is not to be recommended as a routine practice since, as we have already seen, where accurate reproduction is the aim, inspection of the end point of development is not an accurate method. If, however, maximum contrast is required, as it often is in scientific work, inspection is useful. The desensitizer solution should not contain dissolved air, otherwise air bells may be formed and remain adhering to the emulsion during desensitization. They protect the emulsion locally from desensitization and development. Or, if they are washed away by the developer, fog discs are formed when the light is put on during development.

Development and developers

Development is, chemically, a reduction of photo-sensitized silver halide to metallic silver. The modifications in the silver halide, necessary for photography, have been described. Molecular or atomic silver deposited by such a reaction would ordinarily first form a uniform layer. Aggregation and particle growth would then

[1] This dye stains the skin and especially the nails very persistently.

[2] See p. 176 for directions for strength of solutions.

take place around nuclei such as occasional larger groups or foreign bodies. Nuclei of silver or silver sulphide are, however, already present in the photographic emulsion, so that direct deposition of silver occurs on them. If, however, the development is vigorous the resultant silver grain may be considerably larger than the original silver halide crystals which, as Rabinowitsch describes it, are developed explosively. This is evaded by diluting the developer.

The rate of development depends upon the activity of the developing agent (p. 123) and also upon its rate of diffusion through the gelatine. If a very dilute solution of developer is used, its rate of penetration may be so slow that the full density differences corresponding with increased exposure are not reached. By the time that all the latent image in a weakly exposed part of the negative is fully developed, little more may be developed in parts much more heavily exposed, since the same amount of developer will have reached the latent image in these parts and will be insufficient to develop the whole image. Fig. 1, Chapter I, was developed in this manner to equalize the differences of sensitivity of the emulsion to different regions of the spectrum. A useful application of this method for treating negatives, known to have very excessive contrasts, consists of developing them until the image appears in the heavily exposed parts and then transferring the plate without washing into a dish of water. It is then left there for some time. In the weakly exposed parts, there is still left enough developer to bring up the image, whereas, in the heavily exposed parts, it is exhausted and no further development takes place.

A developing solution contains the following components:

1. Reducing agent.
2. Accelerator.
3. Restrainer.
4. Preserver.

The reducing agent is usually an organic reducing agent of one of the types shown below; that is, a di-substituted benzene derivative in which the substituents are two hydroxy-groups, two amino-, or one of each. Further complication is introduced in some cases, usually to increase the vigour of the substance. A halogen atom

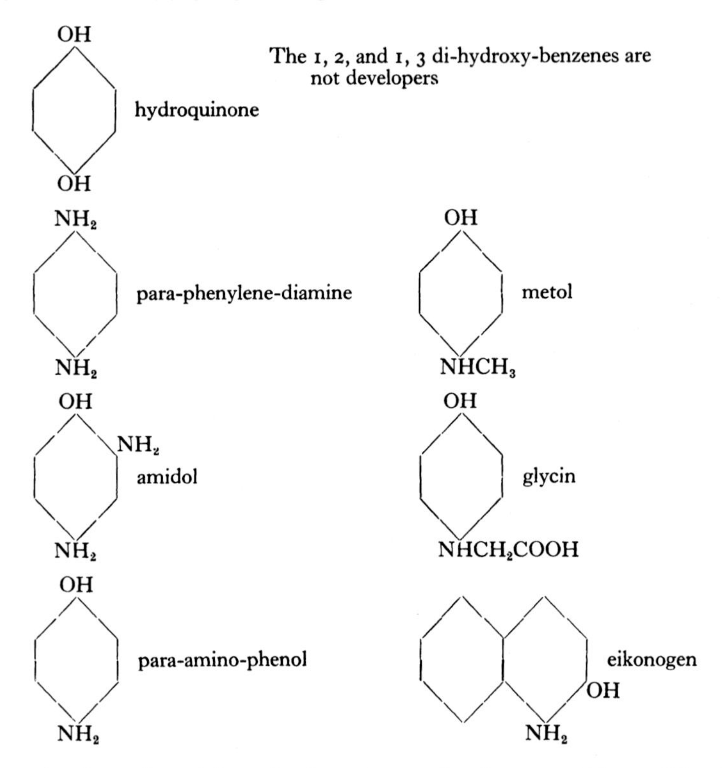

may be introduced into hydroquinone in the 2 position. Metol and glycin are further examples of additional substitution. Pyrogallol, which was formerly used almost exclusively, is 1, 2, 3, trihydroxy-benzene. Its isomers were not used. It is not used now on account of its staining properties.

Most of these reducing agents act only in alkaline solution. When caustic soda is added to them, a phenolate is formed. This property is utilized in preparing concentrated liquid developers which consist of a strong solution of the sodium compound. Carbonate of soda is generally used as accelerator. It avoids difficulties due to caustic soda—ease of attack by atmospheric CO_2 and its decomposing effect on gelatine. The alkali also neutralizes the HBr formed by reduction of the AgBr.

It is now necessary to add potassium bromide as the restrainer to prevent general reduction and chemical fog, since the alkaline developing solution without bromide reduces all the silver, whether exposed or not. Ferrous oxalate, alone of developers, does not require any restrainer.

Fig. 63 shows the effect of soluble bromide upon development. It is characteristic that the effect of the bromide may be expressed as the value δ, which is the perpendicular distance below the exposure axis at which the bromide curves intercept.

The amount of the preserver which also acts as a silver solvent—almost invariably sodium sulphite—depends upon the tendency of the reducing agent to oxidize spontaneously. Glycin requires little sulphite but pyrogallol needs a large amount. Developers used immediately on preparation need only a small amount of sulphite, 1 per cent being sufficient. This concentration is the optimum from the point of view of freedom from chemical fog due to the solvent action of the sulphite. When, however, the solution is to be kept for any time, the amount must be increased.

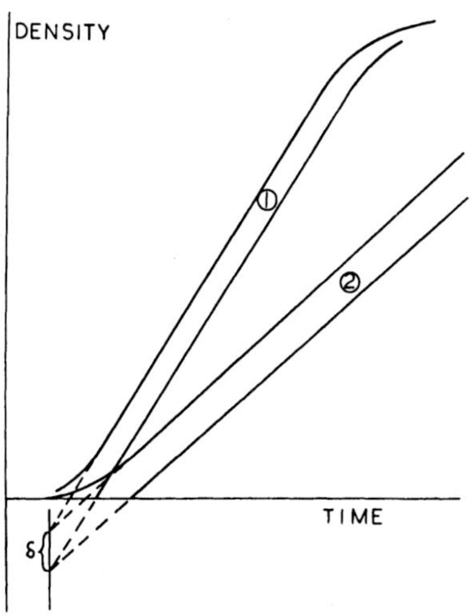

Fig. 63. Effect of bromide upon development. The lower of each pair of curves (different development time) has added bromide.

The spontaneous oxidation of any developer depends upon the amount of atmospheric oxygen available. With rubber bungs this is limited to the amount of air in the bottle. For this reason, bottles should always be kept as nearly full as possible. Where only one supply bottle is kept, this is impossible; but it must be recognized that a small amount of developer at the bottom of a large bottle cannot be kept without serious decomposition.

Choice of a developer

Developing agents have their own special characteristics. Hydroquinone gives strong contrast. For this reason it is used for process plates in copying line diagrams and for similar work. It has the drawback that at low temperatures it becomes almost inactive. It should never be used below 55° F. With metol, shadow detail appears very rapidly. Metol and hydroquinone are therefore excellent together, since metol is a fast soft-working developer while the hydroquinone is complementary and builds up density.[1] Amidol is peculiar in several respects. It develops without addition of alkali, in the presence of sodium sulphite. If sufficiently acid for presence of free sulphurous acid, development does not start at the surface of the emulsion but at the depth below it where diffusion through the gelatine has removed the free SO_2 sufficiently for development to occur. The rate increases as the acidity falls off. Amidol is useful for printing positives because of the very pure black images which it forms.

Temperature coefficient of rate of development

The chemical reaction of development of latent image to metallic silver naturally goes faster as the temperature rises. The temperature coefficient is usually expressed for practical photographic purposes as the ratio between times of development required to give equal results for 10° C. (18° F.) differences. Its variation with different emulsions may be ignored. It is not affected by dilution of the developing bath although variation in the amount of bromide may alter it. The amount of bromide used in the commoner developers has the effect of stabilizing the temperature coefficient, independently of the emulsion, at about 2·0. The following table gives the values for the commoner developers:

Metol	1·3
Paraminophenol	1·5
Ferrous oxalate	1·7
Pyrogallol	1·9
Metol-hydroquinone	1·9
Hydroquinone	2·5
Glycin	2·3
Rodinal	1·9

Once the proper time of development has been found at one temperature, the times required at other temperatures may be calculated knowing that log of development time is a linear function of temperature. It should be noted that calculating times required for given temperatures is a much more reliable method than always trying to work at a fixed temperature by warming solution in cold

[1] Usually referred to as "MQ" developer.

weather, since the temperature of the developing solution will not remain constant during development. It cannot be too strongly urged that the temperature variation of time of development is a serious factor in development technique. M.Q. developers should not be used very cold as the hydroquinone becomes entirely inactive.

Energy of developers

The table below shows the relative energy of some developers, hydroquinone being taken as unity:

Ferrous oxalate	0·3
p-Phenylene diamine hydrochloride (no alkali)	0·3
p-Phenylene diamine hydrochloride (with standard alkali)	0·4
Hydroquinone	1·0
p-Phenyl glycine (glycin)	1·6
p-Aminophenol hydrochloride	6·0
Monomethyl *p*-aminophenol sulphate (metol)	20·0
Diaminophenol (amidol)	30–40

It should be noted that γ_{max} is not a function of the energy nor is the amount of fog formed. Density$_{\infty}$ with paraphenylene diamine developer is 1·7 and γ_{∞} 0·58, compared with values of 3·5 to 4·0 and 1·2 to 1·7 respectively for the ordinary developers.

Technique of development

The simplest method of development is by inspection. In the past this was almost exclusively used and the older workers made much of using their experience to stop development at the correct point, thereby controlling errors of exposure. This method is quite unsound. Variation of the time of development does not compensate for errors of exposure and, as already pointed out, it is desirable to develop always to some fixed γ. Inspection is therefore pointless. The advent of panchromatic materials of high speed has made inspection difficult and the method is used only for slow-speed material. Where accuracy of tone reproduction is not required, as in copying diagrams where maximum contrast is the object, the inspection method can be used with advantage. Incidentally the older workers usually judged completion of development by the appearance of the negative *from the back*. This leads to rather dense negatives, and the results with modern double-coated emulsions are quite impossible. In general, development may be regarded as complete when the whole of the front of the negative is no longer white, provided exposure is correct. The shadows should be faintly grey by the safelight as compared with the clear white of the rebate round the negative. Badly under-exposed parts, which include deep shadows in correctly exposed negatives, will never darken except by fog.

An improvement of the inspection method, which had a certain amount of popularity, was the Watkins factorial development. This depends upon the fact that the time of appearance of the image is an indication of the time required for completion of development. The Watkins factor W is given by the equation $T_c = WT_a$, where T_c is time for complete development and T_a time of appearance of image. W varies according to the developer used and the amount of potassium bromide. The method introduces error if the exposure is not correct, since with under-exposure T_a is long and so the negative is over-developed at T_c and an excessively contrasty negative is obtained. Conversely with over-exposure T_a is small and at T_c a flat negative will be obtained.

For medium fine grain it is convenient to work with developers more dilute than those ordinarily used for dish development. The time of development will thereby increase, and films are most conveniently developed in some sort of tank. Films may be developed in a dish, but the continuous movement of the film through the developer is liable to cause scratches and markings. In the tank all this is obviated. The Zeiss tank is moulded with spirals on top and bottom to hold the film in position. The Correx uses a celluloid corrugated apron, upon which the film is wound, to allow free access of developer. The liquid should always be kept in motion.

Many workers rock their dishes so regularly that stationary waves and concentration differences occur. However, the alternation of dense and less dense strips across the negative needs precision measurements of density to detect it. A similar effect may be seen in the fixing bath just before the last traces of cloudiness disappear.

Fine-grain developers

Since the advent of the miniature (35 mm.) negative and similar small sizes, grain has become a serious factor. At the large magnifications necessary for printing from these small negatives, the silver image is no longer continuous to the eye and the silver grains can be seen separately. With an ordinary strength developer of the M.Q. type, the vigorous action causes the size of the developed grain to spread considerably beyond the limits of the original silver halide crystals. Grain size can therefore be reduced by diluting the developer and keeping the time of development short. Obviously, this should not be overdone, since it is the largest crystals which are the most sensitive to light. Under-exposed negatives therefore show grain worse than fully exposed ones. For very fine grain paraphenylene diamine is the best developer[1] and is 3 to 5 times as efficient in this respect as the ordinary ones. However, it requires extra exposure of 2 or 3 times. Microphotographs are available of the grains in developed films, but these pictures give little information of what we really want to know, namely how sharp is the image of our object. Test enlargements are perhaps the best test so far available. It may be noted that there is possibly a lower optimum grain size since, when it becomes very small, Rayleigh scattering will occur and cause spreading of the image.

A number of medium fine-grain developers have been proposed in recent years, of which the most popular is one containing metol and hydroquinone with a lot of sulphite but no carbonate. Activation of the developer is brought about by a small amount of borax. The grain with such developer is only slightly finer than that obtained by use of normal M.Q. developer diluted to one-quarter, provided that over-development is avoided. Eight times magnification shows no deterioration of definition due to grain.

For miniature negatives, which may require a higher degree of enlargement, the true fine-grain developers are useful. Of these paraphenylene diamine is undoubtedly the most efficient, but it has the drawback that extra exposure of 3 to 2 times must be given. For all ordinary work in daylight this is of course not serious, but for high-speed work in artificial light it may be a drawback. In this case it is better to sacrifice some grain quality and use ordinary M.Q. diluted to one-quarter.

Physical development

It was pointed out as early as 1858 that the latent image is not destroyed by fixation of the plate before development and that, after fixation, silver can be deposited from suitable solutions on it.

[1] See p. 169 for solutions.

Recently, the method has received attention owing to claims made that the resulting grain is very fine. These claims do not seem to have been fully substantiated and the grain is no finer than (if as fine as) that given by the fine-grain developers such as paraphenylene-diamine. Physical development is also rather inconvenient owing to the very long time required.

Solutions are given on p. 169. Chemically clean dishes must be used, otherwise the silver is deposited on the dish and not on the plate. Development time is a matter of hours, and the solution is exhausted inside an hour so that it needs to be renewed. The image is light grey.

Fixing and washing

At the end of development the negative should be washed for a few seconds in running water and then placed in the fixing solution. This washing removes most of the developer, which may otherwise cause staining, and so helps to keep the fixing bath clean. Fig. 64 shows rate of fixation as a function of concentration of hypo. It is obvious, therefore, that the optimum strength of the fixing bath lies between 20 and 40 per cent. It is usual to add some metabisulphite with the object of neutralizing the alkaline developer in the emulsion and thereby stopping development. The scientific accuracy of this belief is obviously doubtful, owing to the time taken by the acid to penetrate into the emulsion, but the presence of metabisulphite is useful in preventing deposits on the film. It may be noted that the action of hypo upon the unchanged silver halide is formation of a double salt: $Ag_2Na_4(S_2O_3)_2$, which then diffuses out.

Washing time varies according to the manner in which the water is caused to flow over the film or plate. Tanks of the type described on p. 124 are good for washing, since the water can reach the whole of the film. Long narrow dishes can be bought for the various sizes of roll film so that they are clipped to each end and then developed, fixed and washed while extended at full length. In washing, ordinary intelligence must be used; air bubbles from a fast supply of water may be a nuisance. Water falling on to a deep tank may not wash properly the bottom of the film, since the hypo solution being denser than water tends to collect at the bottom. A simple method of testing completion of fixation is to take the negative out of the water and allow the drippings from it to fall into a very dilute solution of potassium permanganate. Any hypo present will discharge the pale pink colour. Hypo eliminators, so called, are sold, but since they merely substitute some other substance for the hypo they are not to be recommended.

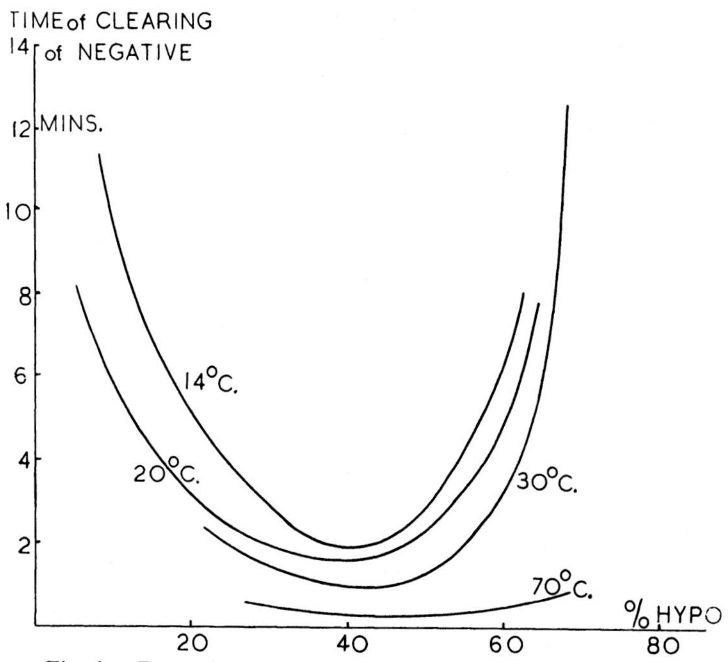

Fig. 64. Dependence of rate of fixing on concentration of hypo.

Drying the film should be carried out in a dust-free atmosphere. Free flow of dry air is much more important than heat. In fact, placing a negative in a hot place without free air supply merely softens the gelatine without hastening drying. Under these conditions one would expect grain growth.

In hot weather, especially, it is useful to use, instead of the ordinary fixing bath, a combined fixing and hardening solution. Addition of chrome alum or of ordinary alum and acetic acid is sufficient; solutions are given on p. 171. Routine use of a hardening bath is particularly useful in the case of miniature films, since it protects them from abrasion and scratching. Hardening also hastens the drying by reducing the swelling of the gelatine.

After drying, negatives should be stored in thin transparent envelopes. This is of the very greatest importance, since negatives stored in bulk rub against each other and in a short time become covered with a multitude of tiny scratches.

Intensification

If the negative, after washing and drying, is found to be inconveniently thin, it may be intensified by a number of methods. It must be noted that a badly under-exposed negative will not be improved by intensification, since obviously the parts of the negative which are blank will remain blank after intensification and only excessive contrast can result. In a few cases, such as when copying black and white drawings and where it is desired to obtain excessive contrast, this does not matter.

The oldest method of intensification consists of bleaching in a solution of mercury chloride whereby the silver is converted into the double mercurous silver chloride. At this stage the negative becomes white; it is washed, preferably with a little acid in the water, and then darkened by one of several solutions. Ammonia blackens the white print but also dissolves out most of the silver chloride. The black image left is not permanently stable. Sodium sulphide solution forms a black image containing half the original silver and a quarter of the added mercury. This image is stable. The most satisfactory method is to redevelop in an organic developer. In this case most of the silver and the mercury are retained in the final image.

Intensification can be carried out in a single bath by use of the mercury-iodide solution (see p. 173). The negative is placed in the solution and left until sufficiently dense.

Chemical reactions taking place are

$$2HgI_2 + Ag_2 = Hg_2I_2 + 2AgI,$$

$$Hg_2I_2 + 2Na_2SO_3 = Hg + HgI_2, 2Na_2SO_3.$$

The intensified image is not quite stable and becomes yellowish after a time; permanent stability, however, is obtained by treating the intensified negative with ordinary developer or with a 1 per cent solution of sodium sulphide. The action of this intensifier is greatest on the lowest densities, a point which makes it particularly suitable for under-exposed negatives. It has a further advantage that intensification may be carried out immediately on removal of the negative from the fixing bath.

Very great intensification is possible with cupric bromide, the silver being converted into bromide which fixes an equivalent of cuprous bromide. The negative after washing is transferred to a bath of silver nitrate. The cuprous bromide reduces the silver nitrate and thus precipitates silver and silver bromide. After washing and reducing the silver bromide there will be three times as much silver in the negative as there was at the beginning.

$$(1)\quad CuBr_2 + Ag = CuBr + AgBr,$$

$$(2)\quad CuBr + 2AgNO_3 \rightarrow Cu(NO_3)_2 + Ag + AgBr.$$

Reduction

Reducing solutions fall into three classes: (1) surface reducers, (2) proportional reducers, and (3) superproportional reducers.

Surface reducers are active substances which attack the silver almost as fast as they penetrate into the emulsions. This type of reducer is useful for clearing fogged negatives. A proportional reducer is one which reduces the density of all the tones proportionately. A good proportional reducing solution consists of potassium permanganate and ammonium persulphate (for quantities see p. 172). It is obvious that the use of proportionate reducers is limited, since the only gain is in shortening the time of exposure in printing. The most useful are the superproportional reducers, since they act on the densest parts of the negative and can be used to correct bad over-development. Ammonium persulphate is the only substance commonly used, usually mixed with sulphuric acid (see p. 172). It should be noted that the persulphate is $(NH_4)_2S_2O_8$ and not the acid sulphate (or bisulphate) NH_4HSO_4.

An ingenious method of reducing density, proportionately in fully exposed negatives and superproportionally in under-exposed and over-developed negatives, consists of toning the image blue by means of ferricyanide bleaching followed by development in ferrous salt solution. This actually intensifies, but the blue image transmits actinic light so that the effect is a reduction.

A convenient reducing solution for slightly veiled prints consists of iodine and potassium cyanide (see p. 173); note that this solution is extremely poisonous. For local reduction of prints a preparation of iodine is on the market.

Faults in negatives

Inexperienced workers usually leave fingerprint marks all over their negatives, both before and after exposure and development. This is easily obviated by always holding negatives by their edges. If the developed negative is dark all over it is due to over-exposure or fog. If the rebate is light and the remainder black, then over-exposure or exposure to light while in the plate holder is the cause; if the rebate is as dark as the rest of the negative, then fog due to exposure to actinic light during development, to stale plates or to bad developer is the cause. Leaky negative holders may cause fog in fairly well defined patches or streaks. Light may also leak in through parts of the camera and form patches of extra density according to the size of the hole. If these are suspected the simplest way of testing the camera is to put a small lamp inside it in the dark room. Any light leaks are then easily observed. If the negative is too thin, it may be due to under-exposure or to loss of power of the developer. The two are easily distinguished, since under-exposure shows itself as large blank patches with the better illuminated portions grey while weak developer produces a uniformly weak image. Density differences with sharp boundaries are due to unequal wetting, so that development has started at different times in different parts of the negative. Air bubbles cause clear discs to appear on the negative, since they prevent the developer from reaching the emulsion at all.[1]

White specks on the negative are due to dust. Plates should be treated with the greatest care to avoid these. Plate holders should be brushed frequently with a fine camel-hair brush. The plates may also be brushed when placed in the plate holder. The camera should also be kept as free from dust as possible. Other common faults are frilling of the emulsion at the edge of the negative and reticulation. These are due to the temperature of the developing bath being too high. This may also cause the negative to appear in low relief on drying. If the negative becomes opaque on keeping, the cause is incomplete washing. Darkening of the negative in patches

[1] Do not wash the plate in tap water before immersing in developer. This is a frequent source of air bells, especially in hot weather.

is due to incomplete fixing. Fog on the negative can be due to age, to chemical fog in development or to light admitted during its time in the camera. In the first two cases, the fog covers the whole plate, while the rebates are clear if the fault occurred while these were protected.

Partial reversal of a negative to a positive may occur when the negative material has been exposed to light at some time—usually during storage. It may also be caused by undue exposure to a dark-room light which is not safe.

On roll films, markings are quite often found extending the whole length of the film as parallel lines. From their appearance they have received the name of 'tram lines'. They are due to grit or roughness on the rollers over which the film is wound. The source of this fault is easily detected and removed.

Lack of sharp definition in the negative may be due to a number of causes. If the focusing is at fault the fuzziness is partial and some planes of the object will probably be quite sharp. If the focusing is grossly incorrect, the fuzziness may be so bad that hardly any detail is perceptible. Fuzziness due to movement of the camera during exposure is almost invariably greater in one direction, on lines at right-angles to that in which the camera was moved. Careful examination of the negative can usually decide to which of these causes lack of sharp definition is due.

Printing processes

Printing includes conversion of negative to positive either by direct contact or by enlarging. Prints may be on paper or glass transparencies for projection in the lantern. Recently use has been made of 35 mm. roll film to prepare series of transparencies for lecture illustration. This method has the defect that order cannot be changed easily nor can additions be made conveniently. Against this, however, the saving of bulk and weight is very great both for storage and for transport. The projector required is also small and simple compared with the large old-fashioned ones used for $3\frac{1}{4}$ inch square glass slides.

For contact printing gaslight paper is almost invariably used. The negative is placed in a wooden frame with the emulsion side in contact with the paper; the back is clipped on and exposure made to an electric-light bulb. The speed of gaslight paper is such that exposures for normal negatives at 2 feet from a 40 watt lamp are of the order of 30 seconds. The slowness of this paper makes manipulation and development[1] in bright yellow light possible.

Bromide paper is considerably faster than gaslight. Its quality is superior and it should always be used where good reproduction of tone values is required. Fig. 65 shows the characteristic curves of bromide papers. Most papers are made in three grades: normal for properly developed negatives, contrasty or vigorous for flat

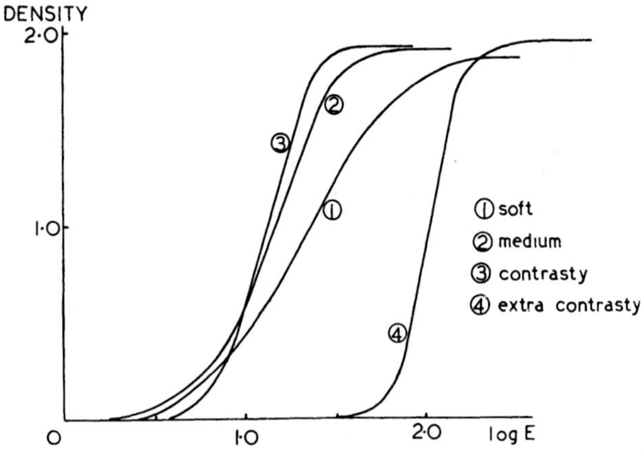

Fig. 65. Characteristic curves of printing papers.

negatives and soft for over-contrasty negatives. It cannot be too strongly emphasized that printing cannot correct serious errors of exposure although it can minimize errors of development. Judgment of the quality of the negatives for choice of grade of printing paper is not made by its overall density or lack of it, but by the range of contrast between high-lights and shadows. For example, an under-exposed fully or over-developed negative is not improved by use of a contrasty paper; quite the reverse. It requires a soft paper. Similarly, a very dense negative, such as commercial developing and printing services usually turn out, does not require a soft paper unless the shadow detail on the negative is absent or very thin. A properly exposed negative will, under these conditions, appear dark all over. It, therefore, requires a normal or even a vigorous paper. A good print almost invariably contains

[1] Developing solution for gaslight paper is given on p. 170.

some parts of the shadows reproduced in the deepest black of which the paper is capable and the highest lights clear white.

Messrs Ilford have prepared the following chart, which shows well the relation between subject range of brightness, negative density range and gamma values:

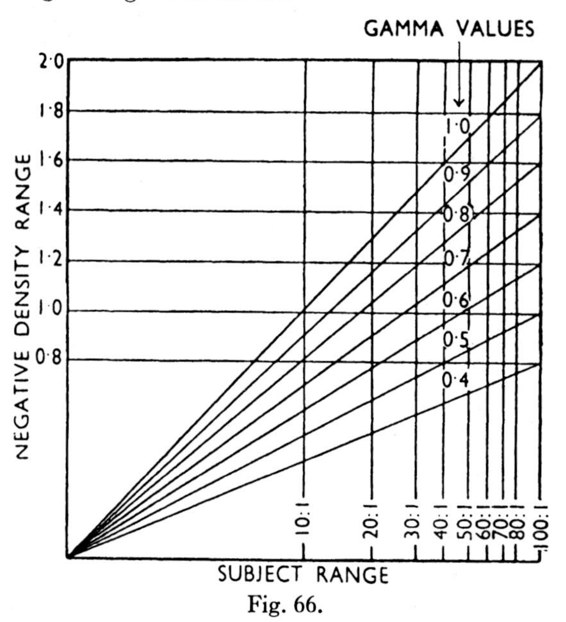

Fig. 66.

The density ranges for which their bromide papers are prepared are as follows:

Grade	Negative density range
Soft	1·8
Normal	1·5
Vigorous	1·3
Contrasty	1·2
Extra-contrasty	1·05
Ultra-contrasty	0·9

This assumes, of course, that it is desired to reproduce the whole of the negative density range on the print.

The technique of printing is very simple once two rules are understood. First, the proper paper must be chosen and, secondly, errors of exposure in printing cannot be rectified by controlling development time without impairing the quality of the print. In making a print, therefore, a test exposure must be made with a small strip of the paper. This is placed in a suitable part of the picture and given exposures over a range which experience suggests, developed and the correct time judged. The whole print is then made with the correct exposure and developed for the proper time. With the developer given on p. 168 this time is 2 minutes at 65° F. Development for less than this time gives greenish-brown tints to the silver image, while increasing the time makes the high-lights muddy. It may be noted that gaslight papers do not suffer from this disability and development may be stopped at any time. Finally, it is very important that prints should be judged in a full bright light; in the dark-room lamp a correctly exposed and developed print always appears too dark, and equally one which appears correct in the dark-room lamp will be underdone when taken out into the daylight. Fixing and washing prints are carried out as described for negatives. Care, however, should be taken that prints do not stick together in the fixing bath or while washing; and that they are not allowed to float to the surface and partially dry there. In the fixing bath it is convenient to place the first print face downwards; the second one is then placed face upwards on it and the two turned over. Safelights should be as bright as possible, which means that the paper should not be unduly exposed to them during development. The simplest way of protecting the paper is to develop it emulsion-side downwards, care being taken to keep the developer moving to prevent formation of air bells.

Enlarging

Fig. 67 shows the set-up of an enlarger. It is shown with a point source of light. This requires that the light shall be moved together with the projecting lens, since the two foci, of condenser and of projecting lens, are conjugate. In practice, this complication is evaded by use of a light source of considerable size, an opal glass gas-filled lamp being most convenient. Lamps approximating to a point source, such as motor-car head lamps, are gradually displacing large-sized high-power lamps in optical projection apparatus. Theoretically this should give better results in enlarging, but the light source must be movable. Enlargers may be bought without a lens and the ordinary camera lens fitted for enlarging.

The following method of testing the lens of an enlarger is simple and accurate. At the centre and corners of an old dark negative sharp scratches are made in the emulsion. The centre is then focused in the enlarger and the amount of stopping down required to make the corners sharp is observed.

In making an enlargement, it is necessary first to make a contact print to decide what part of the picture is required in the enlargement. When this has been decided, the grade of paper to be used

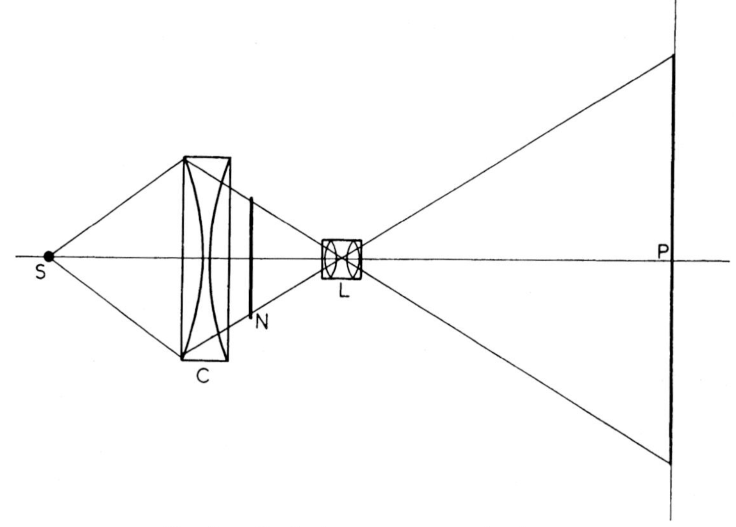

Fig. 67. Optical arrangement of enlarger.

can be chosen from the contrastiness of the part of the negative to be used. This is not always easy, since contrast may be masked by overall density if the negative has been over-exposed. A trial strip of the chosen paper is then exposed so as to receive an important part of the image. From its appearance when developed, it can be decided whether the grade of paper is the correct one. The strip is given a range of exposures by shading and the correct time can be estimated. The use of test strips is particularly important when large prints are being made, owing to the saving in cost. Developing, fixing and washing are then carried out in the manner already described. A useful precaution consists of always wrapping up the unused paper in its black paper as soon as a piece has been taken out. Otherwise one is liable to put the light on to judge the developed print and fog all the unused exposed paper.

After washing, the prints may be freed from surface water by a viscose sponge or by a *clean* towel, then hung up and left to dry. When dry they will be curled; this effect is removed by placing the print face downwards, lifting up one corner and drawing a flat-rigid object, such as a steel rule, from the centre to the opposite corner. This is repeated for the four corners. The print will now be slightly curled in the opposite direction; if placed for a short time under a weight, it will be completely flattened.

Chlorobromide papers are developed by a special developer which gives warm-black tones. A suitable solution is given on p. 170.

Lantern slides are made in two grades, the emulsions corresponding with gaslight and bromide paper. It should be noted that lantern slides require greater density than paper prints, since the range of contrasts possible is much greater and is limited only by the power of the projecting lantern.

In making lantern slides, it is particularly important that results should be uniform: nothing shows up bad dark-room technique more than an over-dark slide followed by one too thin. Obviously, the original negatives need to be correct, and it happens sometimes that a very difficult but important subject is obtained as a thoroughly bad negative. A slide made from this will be bad unless some special treatment is given. Intensification can be used, but it is much better to make a print from the negative, which shows all the required detail. This print is then re-photographed and, by adjustment of the exposure and the development time, a new negative is obtained which will give a satisfactory contact print or slide.

Copying

For copying diagrams, the object to be copied is laid on a base board, held flat by a sheet of plate glass if necessary and illuminated by two or three lamps placed to give even illumination. Where much work of this sort is carried out, a photostat machine is usually used. This is a strongly built camera usually working horizontally and provided with a 90° prism in front of the lens, so that the document to be copied is also horizontal. This prism reverses the image so that copying may be made directly on to paper. In some machines of this type developing and fixing are done in the machine, but this refinement is obviously somewhat dangerous.

a

b

c

Fig. 68. Bellerby pregnancy test—ovulation in Xenopus laevis.
(*a*). Normal lighting (second and fourth from left show ovulation).
(*b*). Through Pola screen.
(*c*). Lighting from beneath.

Specular reflections in record photography

The Pola or other polarizing screen as a sky filter makes use of the fact that some of the light from the sky is polarized by scattering from the numerous particles suspended in the atmosphere. It follows, therefore, that the maximum effect will always be at right-angles to the direction of the Sun's rays. The direction in which the camera must be pointed for maximum effect of the filter is, at sunrise and sunset, an arc from East to West passing overhead and, at noon, the horizon in all directions. In places other than the Equator, where the Sun is not directly overhead at noon, a further correction is needed to make the direction of the camera normal to the Sun's rays. It should be noted that this reasoning does not apply to the use of the Pola screen to cut out specularly reflected light. The consideration upon which this depends is the reflection angle. For non-metallic surfaces light specularly reflected is strongly polarized by the reflection at an angle of about 32°. The Pola screen can therefore cut out reflections from glass only when the light reflected into the camera is all incident from the same direction. With complicated curved surfaces, this condition is not fulfilled and the Pola screen is not much help. When photographing glass apparatus, tanks containing specimens in water and so on, it is best to illuminate from below by placing the object on a sheet of clear or ground glass.

Fig. 68 shows three pictures of the same subject, one, (*a*), with ordinary lighting showing various specular reflections of the light source. (*b*) is with similar illumination but taken through a Pola screen. Most of the light patches have been eliminated but not all. (*c*) is taken with illumination from below (without Pola screen). All patches of light are avoided by this method.

A method sometimes used for softening the specular reflections from polished objects consists of dulling them by dusting them over with fine powder such as talc or, in the case of small objects, cooling them with a little ether or chloroform so that moisture is deposited on them. It is obvious that care must be taken with such methods, since it is desirable that a copy of a polished object should appear polished in the print and not as a matt surface.

CHAPTER VII

SOME SCIENTIFIC APPLICATIONS

It is not proposed to try to collect together in this chapter a representative collection of examples of the use of photography in all the sciences. The most striking results often bear no direct relation to the photographic skill involved, as they are striking because of the subject-matter. Photomicrography provides many examples of subjects which, with quite mediocre or even bad technique, give photographs much more striking than examples of great scientific or photographic importance. The object of this book is to deal with principles rather than working details and the worker is expected to be able to decide for himself which factors must be taken into account in any particular job. It cannot be emphasized too strongly that the photographer must decide what he wants on his plate or film before he starts thinking how to record it. The picture must be seen and analysed by the eye (provided of course that visible radiations are being used) before any attempt is made to photograph it. This analysis also includes recognition of unwanted objects which will be recorded unless special steps are taken to eliminate them. Photographic knowledge must be used twice. The photographer must decide what his two-dimensional picture is going to look like and what he wants it to look like. He can do this only if he knows what photography is capable of doing. Then, having decided what picture he wants, he uses his knowledge of photography a second time to get the result which he has visualized.

There is still a difficulty which may confront both skilled and unskilled photographers in this type of work: what emulsion to use for any given purpose. For ordinary record work, this has been discussed. For some work it is necessary to pick an emulsion with special spectral sensitivity or with other special properties. It is impossible to list here all the characteristics of all the emulsions on the market, and any attempt to do so would probably be out of date by the time it appeared in print. Fortunately, the leading manufacturers are willing and even anxious to help scientific workers in this direction. They should therefore be consulted in case of doubt.[1] During the last few years, there has been a great advance in the supply of technical data made available by the makers of films, plates and papers. Pamphlets can be obtained free or for a few pence from all the reputable photographic dealers.

The subject-matter of this chapter is conveniently divided into three sections:

1. Ordinary photography, mainly straight recording work.
2. Photography by radiations outside the visible spectrum.
3. Cases where photographic materials are used for recording, but without a camera working on the usual optical principles.

Photometry

At first sight, photography would appear to provide an excellent method for measuring light intensities. There are, however, a number of serious difficulties. Only lights of similar colour can be compared. The relation between light intensity and image density is given by the Characteristic Curve. A standardized development must, of course, be used. A simplification consists of making an image of the light source through a Goldberg wedge which gives a continuous Density-Illumination image range with which images of unknown density may be compared to give the corresponding light intensity directly. Part of this problem, which lies outside the scope of this book, is that of measuring the image densities. The photo-electric cell has, however, made it possible to construct efficient Densitometers.

There still remains a serious and fundamental difficulty. The **Reciprocity Law,** which states that the photochemical reaction is directly proportional to the light energy, i.e. to the product of light intensity and exposure time, is not strictly obeyed and breaks down especially with long exposures to weak light and with intermittent exposure. When we consider the theory of Gurney and Mott, these failures are not unexpected.

Record photography

The commonest type of work is Class 1—simple recording of specimens, apparatus, etc. This is straightforward pictorial photo-

[1] Messrs Ilford have prepared a very valuable list of films and plates and the filters to be used with them for all types of scientific work. The title is *Photography as an Aid to Scientific Work*. (No charge is made for this.)

graphy, which has been discussed in previous chapters. Fig. 69 is an example of first-class recording. It shows the peculiar 'spoon' shape of nails, which is due to anaemia resulting from inadequate take-up of iron. It is important to realize that good records are not obtained automatically by just giving the right exposure and making sure that all the details are sharp. These two factors are essential, but a good record of a subject is also a good picture; that is, one which is easy to look at and which tells its own story. It is unfortunate that the high cost of printing prevents wider use of photographic illustration in scientific journals and books; but a record is always useful for making lantern slides. Incidentally, it is a good plan to photograph all important pieces of apparatus as soon as they are completed in the workshop rather than to wait until they are dirty and corroded after use.

Class 1 can be subdivided into two groups: (*a*) where photography is used as an accurate and convenient method for obtaining permanent records. Such records could be obtained less efficiently either by description or by drawing. In the second group (*b*) fall those things which can be recorded only by photographic methods or so much more efficiently and accurately that there is effectively no choice. Examples are the analysis of very rapid motion, recording distant stars and surveying from the air lands otherwise inaccessible.

Class 1 (*a*) can be further subdivided into cases where the object to be recorded is three-dimensional and those where a two-dimensional image is being recorded. One can go on subdividing into additional classes for ever, but there is no point in so doing, especially as the three original classes are not themselves sharply divided. There is, however, some justification in differentiating between two-dimensional and three-dimensional objects. With the latter, correct recording involves all the factors which have already been mentioned as essential for pictorial photography, such as correct arrangement of objects, suitable lighting and viewpoint. On the other hand, the recording of two-dimensional images, such as in photomicrography, does not require all these factors to be taken into account. It is still essential, however, for the operator to decide what features he requires to reproduce, to decide what his picture is going to be and then to make the picture by the suitable photographic methods.

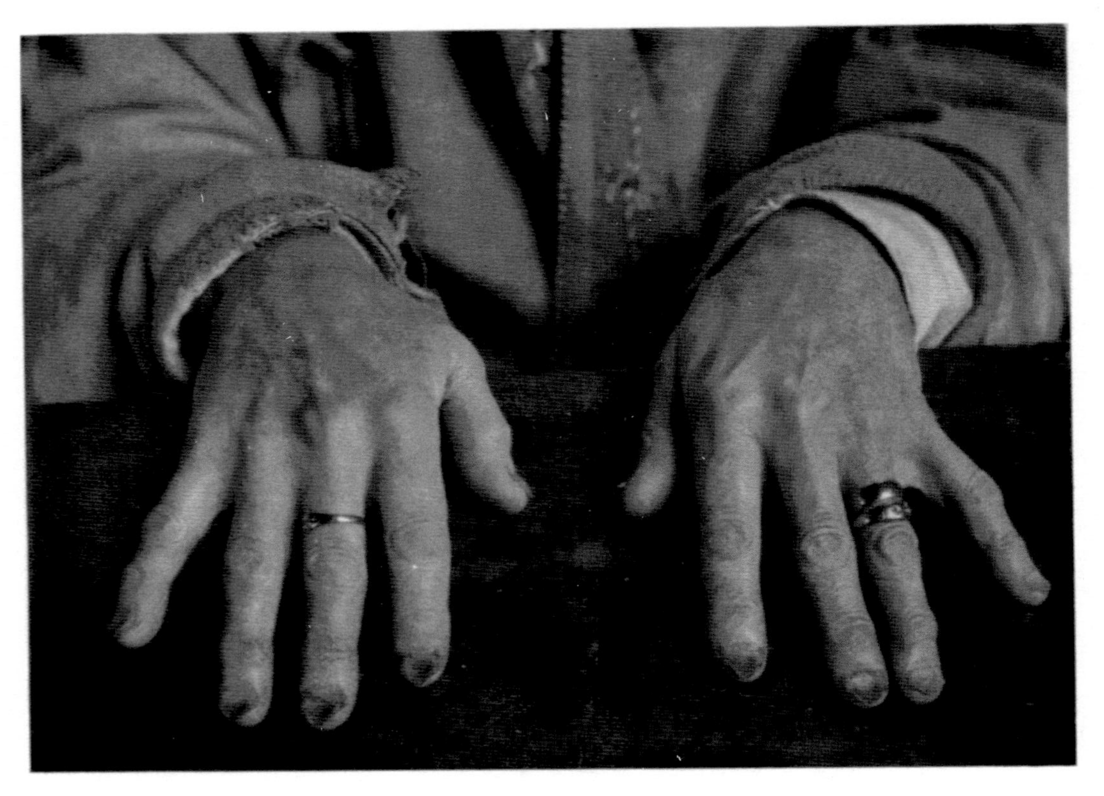

Fig. 69. Medical record photograph.

Fig. 54 (*a*) and (*b*), pages 96, 97, shows an apparatus from two different viewpoints. (*a*) is the orthodox one, showing all the arrangements around the main object and thereby reducing the size of the latter. The direct view gives ugly rectilinear lines although, in this case, their effect upon the pictorial composition is not very bad owing to the redeeming effect of the bench and supporting pillars. Nevertheless, little is seen of the apparatus. In (*b*), however, far more is seen and the viewpoint chosen makes the objects included into a good composition of curves and lines. It not only shows the details required but, by good composition, makes it easy for the eye to see them when looking at the print. I fear that many people would prefer (*a*) to (*b*) because it shows the general lay-out with which the work was done—in fact, for sentimental reasons.

Large-scale photographs of small objects may be obtained without the use of a telephoto lens if sufficient extension can be given to the lens. Using the same symbols as before, i is distance from lens to plate, o is distance from lens to object and f is focal length. M, the ratio of size of image to size of object, is equal to i/o, which is in turn equal to $i\text{-}f/f$. If, then, we express the extension in terms of focal lengths of the lens employed, so that $i/f=n$, then the value of M will be $n-1$. This means that, for reproduction at actual size, the extension must be twice the focal length. This is achieved easily by mounting the lens in a metal tube which is screwed into the lens panel. Magnifications greater than this can be obtained, but as M increases, the depth of focus becomes very small. For recording objects at actual size, or up to twice actual size, this method is very valuable provided the lens can be stopped down to $F/16$ or $F/22$ and the necessarily long exposures given.

It often happens that, owing to limitations of space, the whole of a subject cannot be included in the normal angle of view, α, of the lens. This, as pointed out on p. 32, is taken so that

$$\frac{\frac{1}{2}\text{ diagonal of plate}}{\text{focal length}}=1,\text{ so then }\tan\frac{\alpha}{2}=0{\cdot}5,$$

and α is *c.* 26° and the angle of view *c.* 52°, which is approximately that of the human eye. Lenses including a wider angle than this are called wide-angle lenses. They require a small distance between lens and plate as compared with normal extensions and many cameras cannot be used with such lenses. Field cameras are usually constructed so that the lens can be brought close enough to the plate, either by racking back the lens panel or, occasionally, by racking forward the plate holder. This latter has the advantage that there is less chance of cut-off of part of the rays by the camera base board. In some cameras, the base board is hinged for the same purpose.

Class I (*b*) consists largely of photographic recording of two-dimensional pictures, e.g. the images formed by optical instruments such as the microscope, spectroscope and telescope. It also includes subjects which could not be recorded in any other way, such as objects moving with very high velocity and the analysis of motions where the velocity may be quite moderate but the movement complex. In astronomy, the opposite need occurs—to record, by very long exposure, the image of a star whose luminosity is too small to be detected visually. Photography is also needed by astronomy for recording stars which may be visible through a large telescope but so numerous that any other method of recording is practically impossible.

Another case, which is in a sense also the opposite of the high-speed work already mentioned, is aerial photography. Here, the object is usually stationary but the camera is moving rapidly. Special cameras are made for aerial mapping. Hay (vol. VI)[1] gives a full account of the apparatus used and the methods of analysis of the photographs. The method is especially valuable in tropical or difficult country. Surveys and fixing of frontiers can be carried out in a few days, whereas the same work carried out on the ground would require an army of workers for a long period. Stereoscopic photographs have been taken of Mount Everest from the air. In 1908 a giant meteorite fell in Siberia. Its position has never been determined owing to the difficult nature of the country and climate, but recently the U.S.S.R. have carried out aerial surveys of the region and have prepared a photo-mosaic map which shows the centres of the explosive blast which mowed down avenues of trees.

The use of photography in archaeology, anthropology and geology is largely covered by the remarks on record work in Chapter V. In

[1] *Handbuch der Photographie.*

Fig. 70. Archaeological surveying from the air.

geology, especially, very careful use of filters may be necessary for differentiating between materials which may look identical without the right filter. Aerial photography provides a particularly valuable and simple method of archaeological surveying under certain circumstances. Fig. 70 shows an example which Major G. W. G. Allen, F.S.A., has kindly allowed me to reproduce here. The site, delineated by the differential growth of crops according to the history of the ground on which they grow, is that of a Roman villa. Growth has been retarded where the walls stood. Where the markings are darker, it is due to increased growth over places where ditches have been dug and subsequently filled in. The light region between the villa and the outhouses on the right was found on excavation to have been the threshing floor. When this photograph was taken from the air, nothing of the site was visible on the ground.

High-speed photography

The ordinary iris diaphragm shutter of good make will give exposures as short as $\frac{1}{500}$th of a second. Shorter ones can be given by focal plane blind shutters, but these are limited in two ways. The tension on the blind cannot be increased indefinitely and the usual method of giving short exposures is to use a narrow slit. Often the tension is invariable and the length of the exposure adjusted by altering the width of the slit which passes across the focal plane. This means obviously that, for very short exposures, the slit must be very narrow and distortion occurs in the photograph because the images moving with high velocities are continuously displaced as the slit moves across. The total time to cover the plate is much longer than the effective exposure on any part of it.

When exposures very much shorter are required, a high-tension discharge is used as light source. No shutter is then required, as the duration of the spark is from 10^{-5} to 10^{-7} second according to the electrical circuit used. The intensity of the light from this very short discharge is very great, so that photographs may be taken readily with an ordinary camera. With an induction coil and a Leyden jar, the procedure is simple; the charge builds up in the jar until a discharge takes place. With small sparks from a Leyden jar of small capacity, the exposure is shortest. With larger sparks, the time is considerably longer and is followed by an after-glow lasting for about 10^{-5} second.

The chief difficulty in applying any method for obtaining very short exposures lies in the difficulty of timing the exposure. It is obvious that even with an ordinary camera and an exposure as long as $\frac{1}{500}$th second it will be quite impossible to 'snap' an object which is in the field of view for only the time of the exposure. Fortunately, this is not usually the case but, with really high-speed work, some sort of mechanical release, actuated by the object itself, must be used. Edgerton and others have developed this method lately and have published a number of remarkable results.[1] They charge a condenser to about 2000 volts and this is discharged through the primary of an induction coil to cause a high-tension surge in the secondary. This is discharged through a spark gap or discharge tube. A single picture may be taken or a series. For the latter, the primary circuit is controlled by a commutator on the camera.

High-speed photography is most useful for studying a series of changes occurring very rapidly. This can be done by taking a series of still photographs but it is much more convenient to use a cinematographic method. With the ordinary ciné camera, the film cannot be stopped and started more than about 500 times a second. If the film is allowed to run continuously, it is necessary that the image should move at the same rate as the film on which it is being recorded. One method of evading this difficulty is to use a number of lenses arranged round a disc which rotates in front of the film moving at the same rate. By this means, a few thousand photographs can be taken in a second.

By using a stream of sparks and rotating the film continuously, the number of photographs which can be taken per second seems to be limited only by the rate at which the sparks can be produced. However, with 20,000 sparks per second, the speed at which the film must move becomes a problem. This has been evaded at the Marey Institute by keeping the film motionless, and moving the image optically. The film is mounted around the inside of the circumference of a drum: the image is received by a prism which reflects it on to the film. The small prism is easily rotated at high

[1] *Photogr. J.* 1936, **76**, p. 198. Examples of these have appeared so widely in the popular press that no example is given here.

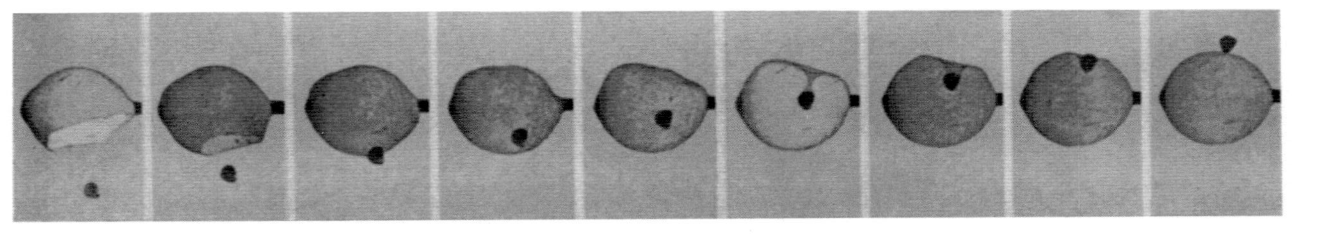

(*a*) Pellet passing through soap bubble.

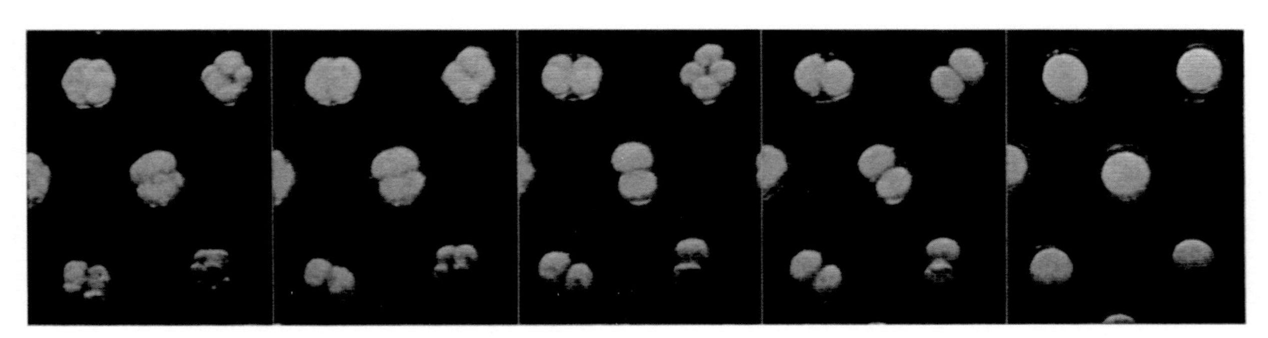

(*b*) Sub-division of sea urchin eggs from the moment of fertilisation. Only every hundredth frame from the original record film is shown here.

Fig. 71. Record cinematography.

speeds, whereas the drum could not be on account of its larger size. Fig. 71 (*a*) shows an example of the results obtained by this method. It is the record of a pellet fired through a soap bubble. This was taken at the Marey Institute.

The timing factor is important with this type of high-speed cinematography for a new reason. If 20,000 pictures are taken in 1 second, nearly 3 miles of film will be used. In the apparatus described, the length of film on the drum is 10 feet, so that the number of pictures $\frac{3}{4}$ inch wide which can be taken in a run is 160. This covers less than $\frac{1}{100}$th sec.

High-speed cinematography has found many uses. Workers doing repetition jobs are photographed and unnecessary movements eliminated. Fatigue is reduced and output increased. The performance of a machine working at high speed can be examined and faults eliminated. Problems as diverse as the removal of metal in screw cutting and the flow of air round aeroplane propellers have been simplified by high-speed photography.

An important use of cinematography, which depends upon the opposite principle to that of high-speed work, is the recording of very slow changes such as the growth of a plant. The time interval between successive photographs may be as long as desired. The continuous speeded-up record, when projected on to the screen, shows in a few minutes changes which may have taken hours or days. Fig. 71 (*b*) shows some frames from a film of this sort illustrating the development of sea urchin eggs.

Photomicrography

Photography is an extremely valuable addition to microscope technique. Permanent and accurate records can be obtained by very simple additional apparatus. Images too feeble for visual examination may be recorded and examined at leisure. Cinematographic records can be made of rapid changes. A full discussion of the microscope lies outside this book, but certain modifications which affect the photographic recording of the image will be dealt with. The numerical aperture of the objective should be as large as possible to reduce diffraction (cf. p. 49). The oil-immersion lens also improves definition by reducing the size of the diffraction image in proportion to the refractive index of the oil. It may be noted that much detail seen in an image is due to differences of refractive index in the specimen. Differences of transparency also occur, of course, and may be the more important in certain cases, such as stained sections.

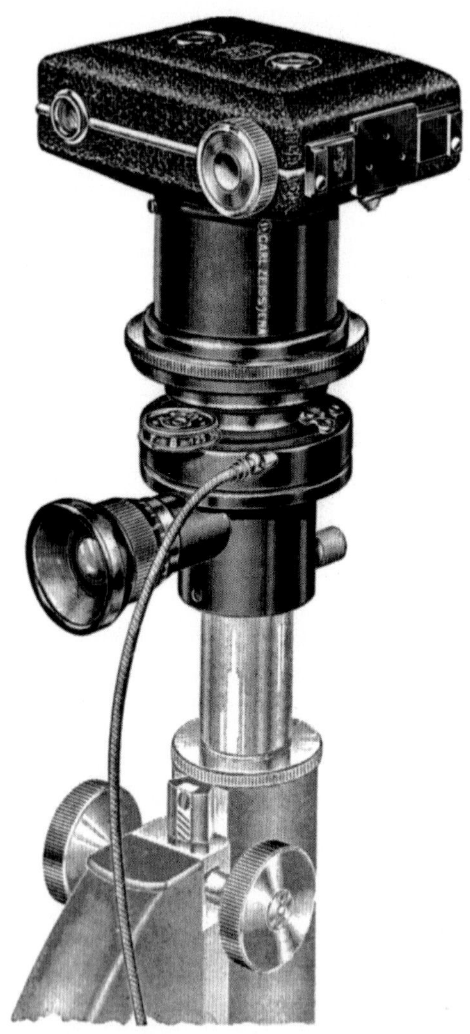

Fig. 72. Camera for photomicrography.

The sole addition that is required to the microscope is a box to hold the sensitive material and some means of focusing the image in the plane of the film or plate. The simplest form consists of a light-tight box attached to the microscope eyepiece by a collar. It is fitted with a simple shutter and antinous release and a ground-glass plate for focusing. Plates are used, usually 4·5 × 6 cm. size. The most serious limitation of this simple apparatus is due to the time required to remove the ground glass, substitute for it the plate holder and to remove its shutter. Also, these operations may shake the apparatus and displace the image. An improved form of the apparatus has a reflex focusing device consisting of a small prism which reflects the light into a side tube through which focusing is done. The image is watched until ready to be exposed. The prism is then pushed out of the field by an antinous release and the exposure made in the usual manner. Fig. 72 shows the form of this apparatus made by Messrs Zeiss.

The commonest fault of photomicrographs is markings due to dirt. Before taking a photograph, all glass surfaces should be carefully examined and cleaned, if necessary. The markings on a photograph due to dirt on the lenses are not easily removed, and it is not always possible to differentiate between them and structure in the object. For this reason, retouching is not allowed.

Photomicrography is subject to all the optical and photographic limitations already described for ordinary photography, and diffraction is especially important at high magnifications. Since diffraction is least for the shorter wave-lengths, a blue-violet filter may be used with advantage. Or a plate sensitive only to the blue-violet end of the spectrum may be employed. Either of these methods also reduces chromatic aberration, which may occur as a primary aberration or as a secondary one due to diffraction. Microscopes have been constructed for use with ultra-violet light to obtain extra resolving power. In these, the optical system is made entirely of quartz.

Halation is always likely to give trouble, especially in pictures of specimens of great contrast. Backed plates should be used, but the best way to reduce this fault is to use film. The apparatus shown in fig. 4 is made for roll film. Many workers prefer plates on account of their flatness. Any gain in this direction is certainly less than the loss of definition due to halation. In the smaller sizes, such as 6 × 4·5 cm. or $3\frac{1}{4} \times 2\frac{1}{4}$ inches, the film may be trusted to be quite flat in any efficiently made camera.

It frequently happens, especially in biological work, that the contrast in the specimen is insufficient to show the details required. This may be due to failure of the sensitive material to differentiate between colour contrasts in the subject or, more usually, because the contrast is not there. Details of structure can then only be shown up by staining. Details which take up the dye can then be shown still more clearly by use of a filter of the complementary colour which transmits the light through the undyed part of the specimen and cuts it off from the dyed part. For example, a specimen dyed pale mauve on a colourless background will give a photograph with little contrast. If, however, this is repeated through a yellow filter, the mauve will show as if black on white.

In photography of coloured objects, such as diffraction images or interference colours, any attempt at correct tone rendering may make the result worse, since the final print will contain a number of grey tones, the original colours of which it is impossible to guess. Such subjects should be taken in monochromatic light, either using a source such as a sodium lamp or by use of filters which transmit a small range of wave-lengths only (e.g. the Ilford spectrum colour filters).

Optical methods for improving visibility in photomicrography

In a number of cases, no details of structure can be seen in a specimen because the field is so flooded with light that small differences of brightness are not perceived. There are two methods by which this difficulty may be overcome. First, by use of an illuminating system which concentrates the light on the object while giving a dark ground. Secondly, a similar effect can be obtained by use of polarized light with polarizing and analysing Nicols crossed. This method, however, works only if the object is birefringent.

The ultra-microscope is the extreme example of the dark-ground method, since it is used on systems in which the particles are too small compared with the wave-length of light to be resolved under any conditions. Fig. 73 shows the principle. Light from an intense source is concentrated, after passing through a slit, in the object under examination (usually a liquid). The concentrated beam is then examined by the microscope at right angles to the direction of the light. When focused, diffraction images around the still invisible particles are visible. Counts may be made of particles in colloidal solutions; their movements may be studied, and the shape of the diffraction image is an indication of the shape of the particles.

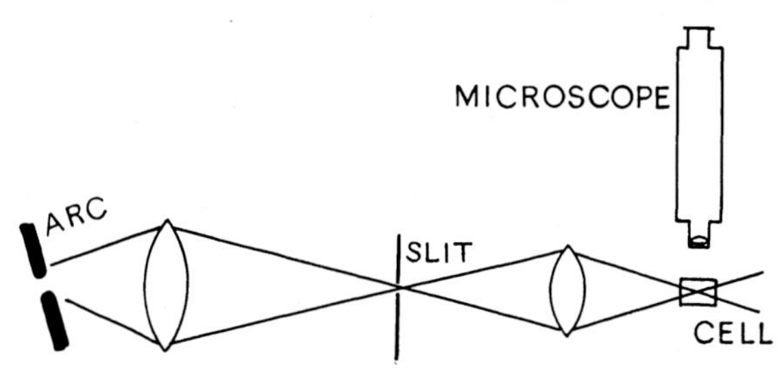

Fig. 73. Optical principle of ultra-microscope.

A simpler form of apparatus utilizing the same principle is the cardioid or dark-ground condenser shown in fig. 74. It has the advantage that it can be attached to any microscope. Light strikes the object under examination only at an angle, all direct light being cut off by a stop. The dark-ground illumination is not reserved only for sub-microscopic particles but can be used in all cases where the dark background is helpful.

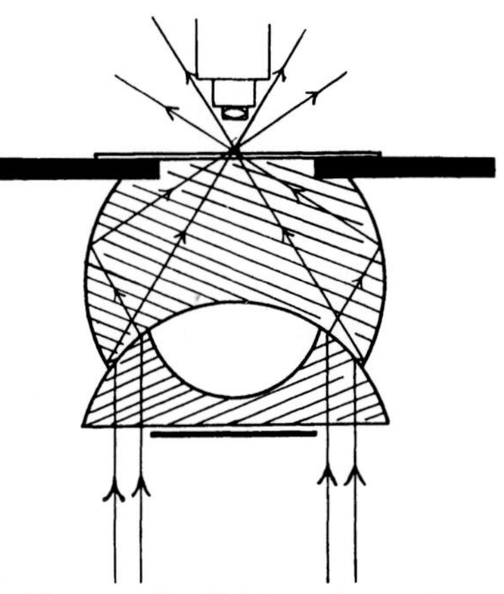

Fig. 74. Cardioid condenser for dark-ground illumination.

An interesting extension of the principle of the cardioid condenser is the Zeiss Micropolychromar, which is a similar arrangement except that light of one colour is used for the indirect illumination of the object while the field is not dark but illuminated by direct light of another contrasting colour. As Messrs Zeiss call it, the system is one of 'optical staining'.

Fig. 75 shows an example of the use of polarized light in elucidation of crystalline structure which cannot be seen clearly by ordinary illumination. (*a*) is by normal illumination and (*b*) by polarized light with Nicols crossed to give a dark background. The crystals (of barium stearate) were very thin.

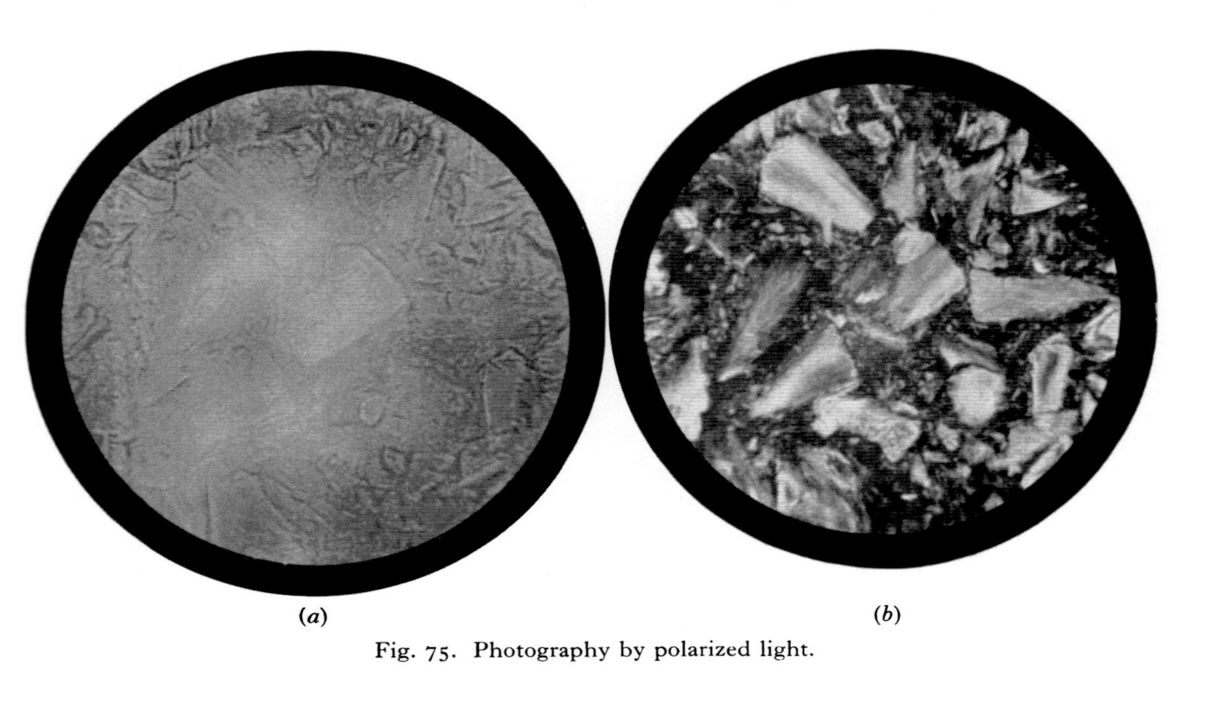

(*a*) (*b*)

Fig. 75. Photography by polarized light.

Formerly, the expense of Nicol prisms of any size was a limit to their use, but the recent introduction of the Pola screen, which can be obtained in large sizes at moderate price, has given an impetus to the use of polarized light. Fig. 76 gives an example of the use of Pola screens for detecting strain birefringence in glass. The screens

Fig. 76. Detection of strain by polarized light.

are crossed to give a dark background. The dark specimen is annealed and free from strain, while the other specimen shows considerable strain.

The Pola screen consists of tiny crystallites of herapathite suspended in nitro-cellulose. They are rod-shaped and are given a common orientation by shearing and mounted between glass plates. No structure can be detected at a magnification of 1100 diameters. When a bright source of light is examined through crossed screens, it is seen as a dull crimson, which is seen through the spectroscope to consist of light from each end of the spectrum. This residual light may be eliminated by a pale green-yellow filter. Other polarizing filters have appeared on the market such as the Bernotar (Zeiss), which consists of a plate of herapathite grown as a single crystal.

Photomicrographs of crystals taken in convergent polarized light and photographs of any other interference pictures should always be by monochromatic light or through a filter transmitting a narrow band of the spectrum. A valuable use of polarized light is due to Prof. Coker, who made models of engineering structures in celluloid and determined the stress distribution by the appearance of birefringence when loads were applied.

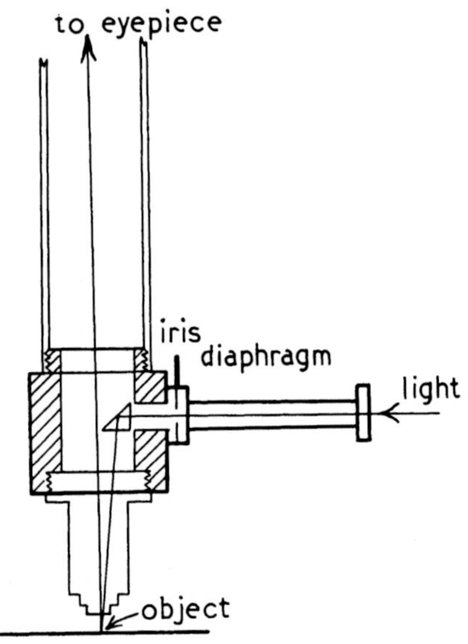

Fig. 77. Principle of reflection microscope.

For examination of opaque objects, illumination is by reflected light, either in a specially constructed microscope or by attaching the apparatus shown in fig. 77 to any ordinary microscope. Light enters as shown and is reflected down through the objective on to the object under examination. The reflected beam then re-traverses the objective in the usual manner. For photography of opaque objects at low magnifications it is sometimes better to use an independent source of light, which can then be directed on the object from any angle desired to show up the texture. Fig. 78 shows an example of the use of this fitting. It is a photograph of the appearance of the 'black spot' in a soap film, showing the two thicknesses of film, the darker being two molecules thick and the next one four. The commonest use of reflection microscopy is for examination of metals. Examples of the results will be found in textbooks on metallurgy. These also describe the methods of preparation of specimens for photographing.

Astronomy

In astronomy, as in microscopy, great magnifications are needed; the largest possible numerical apertures must therefore be used. The limit for telescopes is due to the difficulties of preparing the

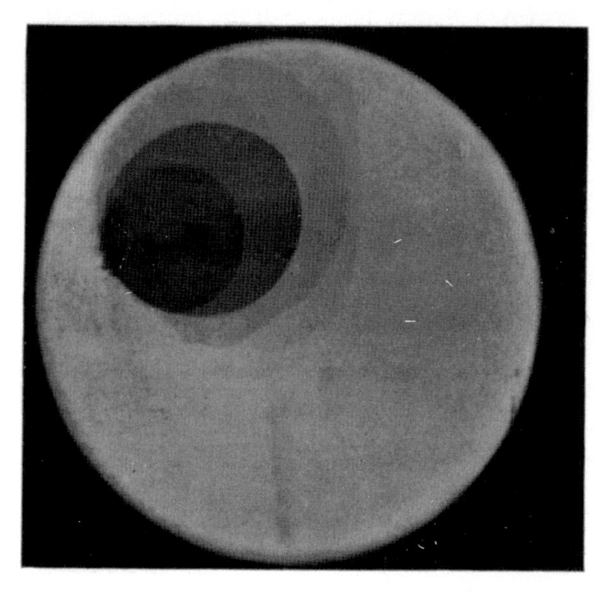

Fig. 78. Growth of "black spot" in soap film. × 120.

necessary large blocks of glass. Astronomical work has one simplifying factor. Depth of focus required is vanishingly small. With a telescope, stars can be detected visually down to the magnitude of about 15. With the aid of photography and long exposures, the 21st magnitude can be recorded. Attempts have been made to photograph very weak images by preliminary short exposure of the plate so that the inertia is already covered before the actual photograph is taken. The advantage for recording very weak images is obvious.

The amount of light reaching an observer on the earth from a distant star will be related to its distance by the inverse square law. Classification of star brightness by magnitudes takes the ratio from one magnitude to the next as 2·512. By this, a 1st magnitude star is 100 times brighter than a 6th magnitude one, since $2{\cdot}512^5 = 100$. Applying the inverse square law, we see that to reduce the apparent magnitude of a star by 1, its distance must be increased by a factor of 1·585, which is the square root of 2·512. It is necessary to distinguish between photographic and visual magnitudes, since the temperature of the star determines what proportion of its radiation lies within the limits of sensitivity of a photographic plate.

The size of the image of a star recorded photographically varies with the exposure according to the equation

$$D = a + b \log E,$$

where D is the actual diameter of the image, a is a variable depending on the conditions and b is called the astro gamma. From the sizes observed with different exposures, the relative brightness of a star can be determined, since E is equal to intensity multiplied by time.

A special method of photographing the Sun by monochromatic light is by the spectroheliograph. An image of the Sun's disc falls upon a slit. The light passing through the slit then passes through a spectroscope arranged so that any chosen wave-length can be admitted to a second slit, behind which is placed the sensitive material. The instrument is moved so that the slit travels past the whole image and produces a composite monochromatic picture of the Sun's disc on the stationary plate.

Another useful application of spectroscopy to astronomy is the utilisation of the Doppler effect to determine the velocity of the star from the displacement of the spectral lines. If a star is approaching the observer with a velocity v, the effect is an apparent reduction of the wave-length of any given line in its spectrum. If V is the velocity of light, λ the wave-length being examined and λ_1 the apparent wave-length recorded, then

$$\lambda - \lambda_1 = \frac{v}{V}\lambda \quad \text{or} \quad v = \frac{V(\lambda - \lambda_1)}{\lambda}.$$

Photography by radiations outside the visible spectrum

Class 2. The first record of the use of photography outside the visible spectrum is Wollaston's discovery of the ultra-violet by its actinic effect. Since then, the extension of the ultra-violet farther and farther has introduced new problems. Most of these are outside photography, but the general problem of absorption is important. Gelatine absorbs, as shown in fig. 8 (*a*), Chapter 1, and the amount in an emulsion must be kept low, as in Ilford *Q* plates, in which the amount is so small that crystals of the silver halide project from the emulsion. For the shortest wave-lengths, Schumann plates are used. These contain no gelatine or absorbing substance and are formed by depositing silver halide from a suspension on to a glass supporting plate. Such 'emulsions' are naturally very fragile. Photography, in the ordinary sense, is not practised with ultra-violet light, although it should be noted that many amateurs unwittingly use it when in the mountains. By neglecting it, they over-expose. The effect is much more serious in colour photography, since it will over-expose the blue-sensitive emulsion: black shadows will appear blue and other colours may be degraded. Ultra-violet light is valuable for microscopy since, according to the equation on p. 51, the shorter the wave-length of the light, the less diffraction will there be. A microscope for such work will need to have its objectives made of quartz, since glass is not sufficiently transparent. It may be noted in this connection that a filter to cut off visible light and let through ultra-violet is provided by a thin layer of silver sputtered on to quartz or a glass transparent to the short wave-lengths; or by a filter of Wood's glass.

Infra-red photography

The deep red end of the spectrum becomes darker as the wave-length increases beyond about 7000. Before 8000 is reached, its visual intensity is so small that there is no clear line of demarcation between deep red and infra-red, just as there is no such separation between the very deep violet and ultra-violet. The upper limit is even more vague, since all wave-lengths beyond the red are 'infra-red' until we reach a region where a new name has been given. This occurs round about 1 million A.U., when we reach the Hertz waves. Infra-red photography is, however, in practice restricted

Fig. 79. Infra-red photograph of the Isle of Wight.

to the region from about 7000 to 13,000 A.U. The best-known application of infra-red photography is to eliminate haze. Photographs have been taken from the air at a distance of more than 300 miles. Such photographs incidentally illustrate visually the curvature of the Earth. The remarkable haze-penetrating properties of infra-red radiations may be explained by the Rayleigh equation (p. 52), which states that the intensity of the scattered light is proportional to the 4th power of the wave-length. Since the blue-violet end of the spectrum, mainly responsible for scattering, has a wave-length of about two-fifths that of the infra-red used, the scattering of the latter will be only $\frac{1}{40}$th of that of the blue-violet.

Fig. 79 shows an example of infra-red aerial photography. It is the Isle of Wight and the South Coast taken from near Poole. It is published by kind permission of *The Times* and Messrs Ilford.

It should be noted that *fogs and mists are not transparent* to infra-red rays. This is probably due to the larger size of the particles of water. By the Rayleigh equation, scattering increases with the square of the particle size, but when the particles reach sizes large compared with the wave-length of visible light, Rayleigh scattering no longer occurs. The opacity of a coarse mist or fog is due to the scattering by internal reflection from drop to drop and to the absorption of light by the drops. Even with the visible spectrum this last factor is of importance, because the droplets are not pure water but are contaminated with soot and the like. Infra-red rays penetrate such fogs slightly or no better than visible light.

There is every reason to suppose that infra-red photography will prove to be of wide value to science when more is known of the relative transparency of substances to these radiations of longer wave-lengths and of their refractive indices for them. It has been shown already that insects which show little detail in visible light are profitably taken by infra-red rays. This is usually ascribed to the transparency of the chitinous layer to these rays. It would appear, however, to be the colouring matter in the chitinous tissue that absorbs the visible light and transmits infra-red. For medical purposes, blood vessels can be photographed through the skin, and skin diseases are seen more clearly than by visible light which shows only surface details. A newly shaven man photographed by infra-red light, looks as if he had omitted to shave because this small penetration below the surface shows the hair roots in the skin.

Recording of X-rays, electrons and other rays

Photography is the most convenient method of recording X-rays and is now used almost exclusively. The use of X-rays for scientific purposes falls into two quite distinct and separate classes:

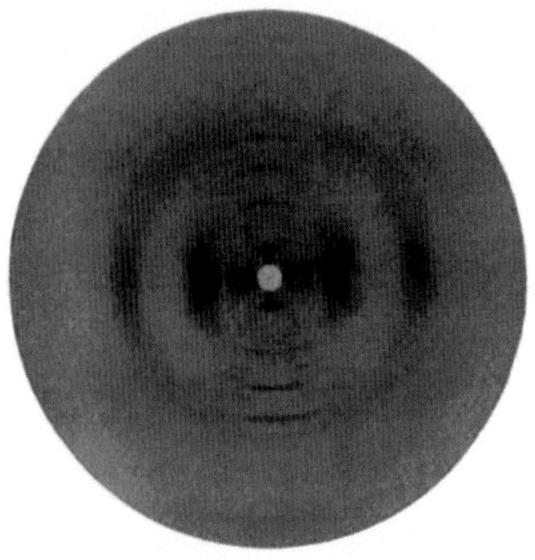

Fig. 80. X-ray diffraction pattern from keratin.

(*a*) Use of selective transparency to X-rays of objects completely opaque to visible light. A silhouette photograph is produced. The opacity of any body depends upon the atomic weight of the elements present. Metal objects in the human body are easily detected owing to their greater atomic weight and opacity. Conversely, faults in metal castings can be detected by their greater transparency. Porous bodies can be photographed by soaking in a solution of a salt of a heavy atom, such as lead, and a silhouette picture is then obtained of the pores. This method has been applied successfully to the examination of the porosity of coal. A similar principle applies to the medical use of a bismuth meal. The patient is given a meal of a harmless compound of bismuth which is opaque to the X-rays.

(*b*) The second use of X-rays depends upon the shortness of their wave-length compared with that of visible light. Interference occurs between the regularly spaced planes of atoms in crystalline substances. Photography is the most convenient method of recording the interference pattern. Spectacular results have been obtained from a large number of crystalline substances and also from many substances not fully crystalline, such as natural and synthetic fibres, rubber and other polymers.[1] Fig. 80, for which I am indebted to Dr W. T. Astbury, is the X-ray diffraction pattern from feather keratin.

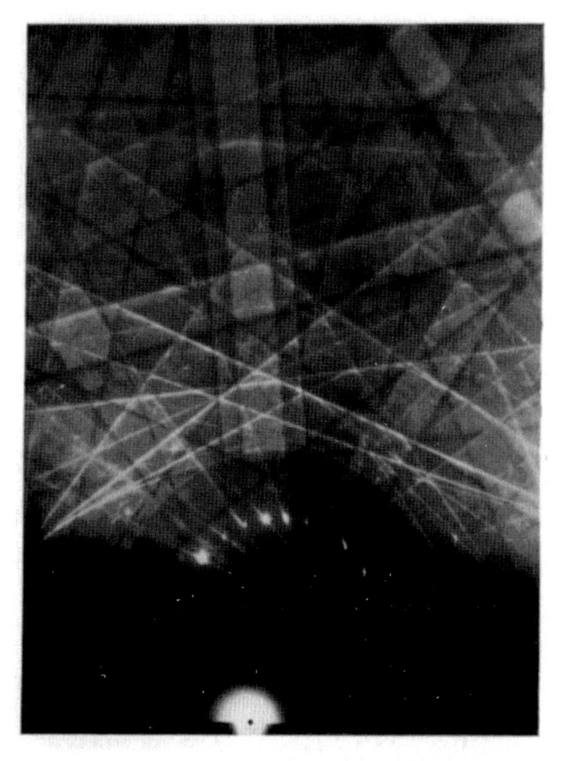

Fig. 81. Electron diffraction pattern from crystal surface.

Electrons can be used for detection of the structure of matter in a similar manner to X-rays. They are less penetrating and are therefore of the greatest value in examination of surface structure. Fig. 81 is a good example, for which I am indebted to Prof. G. I.

[1] *X-rays and Crystal Structure*, by W. H. and W. L. Bragg; *Fundamentals of Fibre Structure*, by W. T. Astbury; *The Diffraction of X-rays and electrons by amorphous solids, liquids and gases*, by J. T. Randall.

Finch. It shows the single crystal electron diffraction pattern given by a fresh cleavage surface of a crystal of calcite. It was taken with a beam of fast electrons at a glancing angle of one or two degrees. The sharpness of the pattern testifies to the perfection of the crystal structure in the surface layers of the crystal.

Photography is also used for recording other 'radiations' of high energy such as alpha rays, protons, neutrons and cosmic rays: and, in the mass spectrograph, positive rays. Originally, the word 'photography' meant the recording of a scene or object by utilizing the actinic effect of visible light upon silver compounds. Many of the uses cited in this chapter are not strictly 'photography' because light is not used and because we are not making a picture of something. The principle, however, by which a radiation makes its mark is the same principle as that by which light is recorded. A latent image is formed and is then developed. It has thus come about that the word 'photography' has extended its meaning. In its original meaning, both pictorial and technical aspects of the subject were included. Attention has been given already to the essential difference between these two aspects of the art and to the difficulty of combining them.

Preparation of prints for publication

Photographs which are to be printed by photo-mechanical methods should be somewhat more contrasty than ordinary prints. Glossy paper is best, since the printer makes a new negative. If, however, it is a picture of apparatus which may need retouching to strengthen the outlines, then a matt surface paper should be used. Kodak Finisher bromide paper is an excellent one for this purpose. It is often stated that no photograph to be printed in a scientific journal should be retouched at all. This is manifestly absurd for two reasons. First, the processes of exposure, development and the subsequent printing all allow a considerable control of what will appear on the final print; more control, in fact, than the ordinary unskilled worker is capable of reaching by manual retouching. Secondly, many photographs, such as those of apparatus, need retouching to show the details desired. Without this retouching, they would not be worth publishing as illustrations. The rule should be that no detail should be eliminated, strengthened or altered in shape when there is any possible doubt about its real shape.

The commonest and most objectionable-looking blemish on a print is the black spot due to a pinhole in the negative. This is also difficult to remove. Of course, the proper method to deal with this fault is to repeat the photograph with more care. If this is impossible, the pinholes may be filled up on the negative. This can be done on the back of the negative support, so that the negative is not itself touched. Alternatively, they can be removed from the print. The simplest method is by gentle scraping by a safety razor blade. This is quite satisfactory for reducing a dead black spot to any shade of grey required, but it is difficult to remove entirely without leaving obvious traces. An alternative method is to use a chemical bleaching solution, such as the potassium ferricyanide-hypo or potassium cyanide-iodine solutions given in the Appendix. The black spot is wetted with water applied by a fine camel-hair brush and the reducing solution is then applied in the same way. The whole print is then thoroughly washed. Care must be taken not to apply too much of the reducing solution, otherwise it will spread beyond the spot to be removed.

White spots on prints may be darkened by wetting the spot and then applying black pigment with a very fine brush until the desired effect is reached.

Conclusion

In conclusion, let me say once again that the photograph obtained is that which the operator should have visualized when he was arranging the object, lighting, viewpoint and so on. A picture of a three-dimensional object has got to be shown in two dimensions, whether made photographically or by any other means; and the two-dimensional picture must give the impression of the three dimensions of the original object. This can be done by use of perspective, directed lighting, depth of focus, etc., as described already, but it must be realized that all this is quite distinct from correct photographic technique, which may be defined as obtaining the effect visualized. If the picture is wrongly visualized, or 'composed' as it is usually termed, correct photographic technique will merely reproduce the bad effect. Composition is aided by experience, but bad composition is more often than not due to failure of the photographer to decide exactly what effect has taken his fancy. In outdoor work, colour is the commonest source of error. The appeal of a scene to the eye may depend upon its colour. The colours are merely tones of grey on the photograph and much of their contrasts is entirely lost. It is of no use to use a so-called correct colour filter on the assumption that 'correct' means giving the effect desired. The colour of the filter must be chosen to give contrast where it is wanted. Often this can be done only by incorrect rendering of the colours.

The rules of photography—both the subjective pictorial ones as well as the more generally recognized ones of technique—are not particularly numerous or complex once the general principles are understood. Without this, photography is a rather wild form of guesswork in which the odds are greatly against the button-pusher. In fact, it is due to the simplicity of photography and to the remarkable latitude of photographic materials that the unskilled amateur gets results even as good as he does.

APPENDIX I

FORMULAE FOR SOLUTIONS

NOTE. *Figures have been reduced in most cases to the amounts required for* 1000 c.c. *of solution*

Developers for plates or films

Pyro-Soda Developer

STOCK SOLUTION

Pyrogallic acid	25 gm.
Potassium metabisulphite	6 ,,
Water up to	250 c.c.

The potassium metabisulphite should be first dissolved in the water previous to adding the pyro.

WORKING SOLUTIONS

A

Stock solution of pyro	50 c.c.
Water up to	500 ,,

B

Sodium carbonate (cryst.)	50 gm.
Sodium sulphite (cryst.)	50 ,,
Potassium bromide (10 per cent solution)	6 c.c.
Water up to	500 ,,

Dish. For use, mix equal parts of A and B.
Tank. ,, ,, 1 part A, 5 parts B and water 20 parts.

Metol-Hydroquinone Ilford

STOCK SOLUTION

Metol	2 gm.
Sodium sulphite (cryst.)	150 ,,
Hydroquinone	8 ,,
Sodium carbonate (cryst.)	100 ,,
Potassium bromide	2 ,,
Water up to	1000 c.c.

Dish. For use, dilute 1 part with 2 parts of water.
Tank. ,, ,, 1 ,, 5 ,, ,,

M.Q. Agfa

Metol	1·07 gm.
Sodium sulphite (anhydrous)	128·6 ,,
Hydroquinone	18 ,,
Sodium carbonate (anhydrous)	46 ,,
Potassium bromide	0·64 ,,
Water up to	1000 c.c.

Developing time: 10–12 min. at 18° C.

Marion M.Q. developer

Potassium metabisulphite	2·5 gm.
Metol	2·5 ,,
Hydroquinone	10 ,,
Potassium bromide	1·5 ,,
Sodium sulphite (cryst.)	150 ,,
Sodium carbonate (cryst.)	150 ,,
Water up to	1 litre

For plates and films this solution is diluted with parts of water. Development then requires 2 minutes at 20° C. For bromide papers, it should be diluted with 3 parts of water. This solution diluted, as described, has been used by the Cambridge University Camera Club for several years as its stock one. It gives grain only very slightly coarser than that given by the M.Q. sulphite-borax medium fine grain solution.

Fine Grain development

Medium Fine Grain: D. 76 (Kodak)

Metol	2 gm.
Hydroquinone	5 ,,
Sodium sulphite anhydrous	100 ,,
Borax	2 ,,
Water up to	1000 c.c.

Real Fine Grain: Sease III

Sodium sulphite anhydrous	90 gm.
Paraphenylene-diamine	10 ,,
Glycin	6 ,,
Water up to	1000 c.c.

Paraphenylene-Diamine-Glycin-Metol
(published in the *Leica Manual*)

Sodium sulphite anhydrous	90 gm.
Paraphenylene-diamine	10 ,,
Glycin	5 ,,
Metol	6 ,,
Water up to	1000 c.c.

Developing times for the above developers using Agfa ISS or Kodak Super X and developing at 20° C.:

D. 76	... ...	6 min. 40 sec.
Sease III	... ...	20 min.
Para-Metol	... ...	16 min.

At 18° C.:

D. 76	... ...	8 min.
Sease III	... ...	25 min.
Para-Metol	... ...	20 min.

Simple Very Fine Grain Developer

Paraphenylene-diamine	13 gm.
Sodium sulphite	65 ,,
Borax	36·5 ,,
Tribasic sodium phosphate	30 ,,
Water up to	1 litre

Development time is 35 minutes at 20° C. The extra fine grain paraphenylene-diamine developers require about three times the ordinary exposure.

X-ray Film Developer

Metol	2 gm.
Sodium sulphite (cryst.)	150 ,,
Hydroquinone	8 ,,
Sodium carbonate (cryst.)	100 ,,
Potassium bromide	5 ,,
Water up to	1000 c.c.

Wash for 1 minute before fixing.

Developer for Contrast, e.g. diagrams on process plates

A

Hydroquinone	25 gm.
Potassium metabisulphite	25 ,,
Potassium bromide	25 ,,
Water up to	1000 c.c.

B

Potassium hydrate (stick)	50 gm.
Water up to	1000 c.c.

For use, mix equal parts of A and B.

With normal exposure development should be complete in about 2 minutes.

Developers for bromide papers

Metol Hydroquinone

Metol	1·5 gm.
Sodium sulphite (cryst.)	50 ,,
Hydroquinone	6 ,,
Sodium carbonate (cryst.)	80 ,,
Potassium bromide	2 ,,
Water up to	1000 c.c.

For use, dilute 1 part with 1 part of water.

Amidol Developer

Sodium sulphite (cryst.)	50 gm.
Amidol	6 ,,
Potassium bromide (10 per cent solution)	8 c.c.
Water up to	1000 c.c.

Development with the above developers should be complete in about 2 minutes. After development rinse and transfer to the fixing bath.

Ilford Clorona Paper

Metol	0·5 gm.
Chlorquinol (or adurol)	6·2 ,,
Hydroquinone	6·2
Sodium sulphite (cryst.)	100 ,,
Sodium carbonate (cryst.)	100 ,,
Potassium bromide	0·8 ,,
Water up to	2000 c.c.

Development should be completed in about 1½ minutes.

One part of this developer mixed with 3 parts of water gives a warm black colour in about 3 minutes. More exposure and dilution with 6 parts of water gives a sepia in about the same time.

Colder or warmer tones may be obtained by considerably reducing or increasing the amount of potassium bromide.

Gaslight Paper Developer

Metol	3 gm.
Sodium sulphite (cryst.)	100 ,,
Hydroquinone	12·5 ,,
Sodium carbonate (cryst.)	187·5 ,,
Potassium bromide	0·75 ,,
Water up to	2000 c.c.

Gaslight prints should be fully developed; then rinsed, fixed and washed as for bromide paper.

Acid Fixing Bath

Sodium hyposulphite	400 gm.
Potassium metabisulphite	25 ,,
Water up to	1000 c.c.

Combined Fixing and Hardening Bath

Sodium hyposulphite	300 gm.
Potassium metabisulphite	25 „
Chrome alum	12·5 „
Water up to	1000 c.c.

The hypo and the metabisulphite are dissolved in 750 c.c. of hot water and allowed to cool. The chrome alum is then dissolved in 150 c.c. of warm water and added to the remainder of the bath when cool.

Fixing Bath for Bromide Papers

Sodium hyposulphite	200 gm.
Potassium metabisulphite	25 „
Water up to	1000 c.c.

Combined Fixing and Hardening Solution for Papers

Some papers are stained by the chrome alum hardener. Prints which are going to be glazed should be fixed in this solution.

A

Hypo	250 gm.
Water up to	1 litre

B

Sodium sulphite	100 gm.
Acetic acid (glacial)	150 c.c.
Alum	100 gm.
Water up to	1 litre

Dissolve the sulphite in 250 c.c. of the water, which may be hot to hasten solution. When the solution is cold, add the acetic acid slowly with constant stirring. The alum is dissolved in 500 c.c. of hot water, cooled and the sulphite-acid solution added to it at a temperature not exceeding 20° C. The remainder of the water is then added to make the bulk up to 1 litre. For use, 2 parts of B are added to 20 parts of A.

Intensifiers

Mercury Intensifier

BLEACHING SOLUTION

Mercuric chloride	27·5 gm.
Water up to	1000 c.c.

When bleaching is complete, wash in water for a few minutes and then in two or three very dilute hydrochloric acid baths. The negative is then washed again in running water for a few minutes and re-developed in one of the following solutions:

A. 5 per cent ammonia.

B. 10 per cent sodium sulphite.

C. Standard metol hydroquinone developer.

Chromium Intensifier

BICHROMATE STOCK SOLUTION

Potassium bichromate	25 gm.
Water up to	250 c.c.

This solution keeps indefinitely.

BLEACHING SOLUTION A

Bichromate stock solution	12·5 c.c.
Hydrochloric acid (conc.)	0·3 ,,
Water up to	125 ,,

BLEACHING SOLUTION B

Bichromate stock solution	12·5 c.c.
Hydrochloric acid (conc.)	1·5 ,,
Water up to	125 ,,

The bleaching solution should be freshly made. Solution A gives more intensification than solution B. Immerse the washed negative in one of these solutions until it is entirely bleached, then wash until the yellow stain is removed from the film, and develop, by daylight or after exposure to daylight, with a negative developer.

Thorough washing is necessary after intensification by any process.

Single Bath Mercury Intensifier

Sodium sulphite	200 gm.
Mercuric iodide	10 ,,
Water up to	1000 c.c.

The sodium sulphite must be dissolved first. This solution keeps well in the dark. The plate can be placed in this bath after a very short wash after fixing.

Cuprous Bromide Intensifier

The negative is bleached in a mixture of the two following solutions:

A

Copper sulphate	23 gm.
Water up to	100 c.c.

B

Potassium bromide	23 gm.
Water up to	100 c.c.

After bleaching, the negative is given a short wash and is then blackened in solution C.

C

Silver nitrate	10 gm.
Water up to	100 c.c.

This method gives very great intensification and is suited for line drawings.

Reducing solutions

Ferricyanide or Farmer's Reducer

Increases contrast by reducing density in shadows much more than in high-lights.

Potassium ferricyanide	2·5 gm.
Water up to	25 c.c.

A fresh plain 20 per cent solution of hypo is used and sufficient ferricyanide solution added to colour the hypo pale yellow. The energy of the reduction is proportional to the amount of ferricyanide present and the process of reduction should be closely watched. Thorough washing is all that is required afterwards.

Persulphate Reducer

Decreases contrast, acting first on the dense high-lights.

Ammonium persulphate	6 gm.
Water up to	250 c.c.

One or two drops of sulphuric acid should be added to induce regularity of action.

Proportional Reducer

Acts proportionately on the densities of the negative.

A

Potassium permanganate	0·12 gm.
Sulphuric acid (conc.)	0·75 c.c.
Water up to	500 ,,

B

Ammonium persulphate	12·5 gm.
Water up to	500 c.c.

For use, mix 1 part of A with 3 parts of B.

Reducer

Stock Iodine Solution

Potassium iodide	6 gm.
Iodine	1 ,,
Water up to	250 c.c.

Stock Cyanide Solution

Potassium cyanide	2 gm.
Water up to	250 c.c.

N.B. Potassium cyanide is a very strong poison and must be used with extreme care. It should be used fresh as the solid decomposes rapidly in air.

For use take 25 c.c. of each stock solution and make up to 500 c.c. with water.

Blue Toning Reducer

Reduces contrast for printing.

BLEACHING SOLUTION A

Potassium ferricyanide	1·7 gm.
Sulphuric acid (conc.)	3 c.c.
Water up to	1000 ,,

BLEACHING SOLUTION B

Ferric ammonium citrate	1·7 gm.
Sulphuric acid (conc.)	3 c.c.
Water up to	1000 ,,

Mix equal parts of A and B when required for use. When toned, wash till all yellow colour is removed.

The Dufaycolor Process

Constant First Development Method

Work in total darkness or with an Ilford 'G' Safelight shielded so that the direct light cannot fall on the film.

Immerse in one of the following developers for 3 minutes at 65° F. (2½ minutes at 70° F. or 2 minutes at 75° F.), vigorously rocking the dish the whole time; agitation increases the brilliance of the transparencies. Formula B gives slightly more contrasty results than A and has the advantage that the thiocyanate is a more constant and stable chemical than ammonia. When using the formula A it is important to ensure that the ammonia is up to its labelled strength.

DEVELOPER A

Metol	6·5 gm.
Sodium sulphite (cryst.)	100 ,,
Hydroquinone	2·0 ,,
Potassium bromide	2·75 ,,
Ammonia (specific gravity ·880)	11 c.c.
Water up to	1000 ,,

DEVELOPER B

Metol	6·5 gm.
Sodium sulphite (cryst.)	100 ,,
Hydroquinone	2·0 ,,
Sodium carbonate (cryst.)	100 ,,
Potassium bromide	2·75 ,,
Potassium thiocyanate (sulphocyanide) (pure)	9 ,,
Water up to	1000 c.c.

It is important that the chemicals should be dissolved in the order given. Alternative ammonia strengths are: Ammonia specific gravity 0·910 = 15 c.c.; Solution of Ammonia B.P. = 34 c.c.

These developers do not keep indefinitely; if a stock solution is to be kept, then formula B is preferable, but it is not advisable to use this if more than two or three weeks old. If old developer is employed rather weak and flat transparencies will be obtained.

It is preferable to use fresh developer for each film unless several films are developed at the same time; the used developer may, however, be employed again for the second development process.

Bleaching Bath

After development rinse the film for about 2 minutes or place immediately in a 1 per cent solution of acetic acid to stop development. Then transfer to the following bath for 5 minutes until all the negative silver image is dissolved.

Potassium permanganate	2 gm.
Sulphuric acid (conc.)	10 c.c.
Water up to	1000 ,,

It is important to agitate the film during the bleaching operation.

In making up this solution it should be remembered that the permanganate may take some hours to dissolve completely and the sulphuric acid should be added a few drops at a time and not suddenly. It is desirable to use water free from any appreciable quantity of chlorides or else a trace of free silver nitrate (0·5 gm.) should be added, as this will prevent the deterioration of the solution during storage. The solution should not be kept after use.

White light may be turned on in the dark room when the film has been in the bleaching bath for about 1 minute, and all subsequent operations carried out in the white light.

Clearing Bath

After bleaching, the film is rinsed and placed in the following bath until the brown stain has disappeared:

Potassium metabisulphite	25 gm.
Water up to	1000 c.c.

Alternative Bleaching Bath

Some workers prefer the bichromate bleach because it is more stable and no clearing bath is required after it, thus obviating one operation. If the bichromate bleach is used the resulting Dufaycolor transparencies will be definitely softer and more degraded in colour. For some purposes these relatively degraded effects may be considered more pleasing and harmonious.

Bleaching requires 3–5 minutes in a bath prepared by diluting 1 part of the following solution with 10 parts of water.

Potassium bichromate	50 gm.
Sulphuric acid (conc.)	100 c.c.
Water up to	1000 ,,

Dissolve the bichromate in the water first and then add the sulphuric acid.

APPENDIX II

The fundamental photographic equation

$$\frac{1}{f}=\frac{1}{i}+\frac{1}{o} \qquad \ldots\ldots(1)$$

is as follows for a thin lens. In fig. 82 the triangles COO_1 and CII_1 are similar so that

$$\frac{II_1}{O_1O}=\frac{CI}{CO}.$$

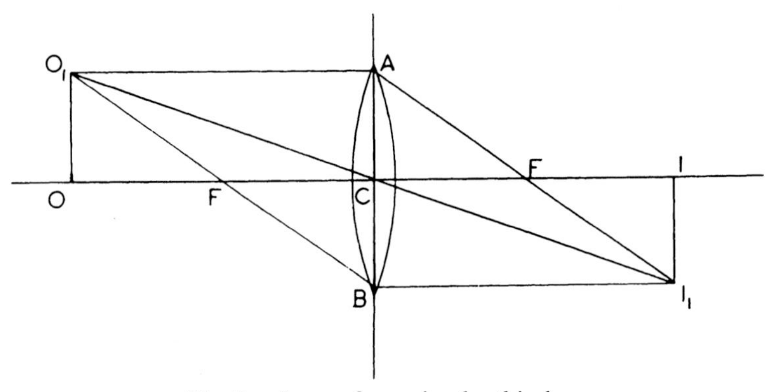

Fig. 82. Image formation by thin lens.

The triangles CAF and FI_1 are also similar, so that

$$\frac{II_1}{AC}=\frac{IF}{FC},$$

but $AC=O_1O$.

Therefore

$$\frac{CI}{CO}=\frac{IF}{FC},$$

which is

$$\frac{i}{o}=\frac{i-f}{f}.$$

This reduces to equation 1. As it is, it gives the magnification—size of image over size of object—in terms of focal length and distance of image from lens. A more useful form of equation 1 gives magnification in terms of object distance and focal length:

$$M=\frac{i}{o}=\frac{f}{o-f}.$$

When o is large compared with f (as it usually is) the magnification becomes f/o.

If the lens is not a thin one, the actual path of the light through it or through a combination of lenses must be considered. Fig. 83 shows (in full lines) the actual path of the light from an object O_1 to its image I_1 and (in broken lines) the Gauss treatment for the ray from O_2 to I_2. N_1 and N_2 are the entry and exit nodal planes. Their position is such that the space between them can be ignored and the simple equations used provided that o and i are measured from the nodal planes to O and I. C is

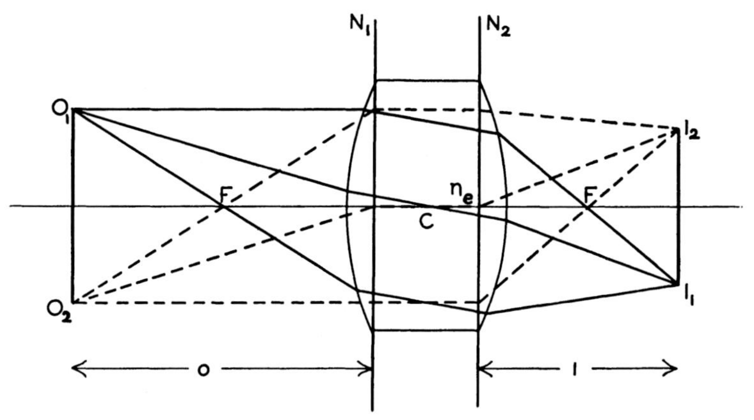

Fig. 83 Image formation by thick lens. Full lines O_1 to I_1 show actual path of lightrays and the broken lines O_2 to I_2 the Gauss treatment.

the optical centre and the diaphragm should be placed as close to this as possible. Since the rays from O_1 and O_2 are parallel to the optic axis, they represent the path of a ray from infinity and the point, F, of intersection with the axis is the focus. The principal focal length of the lens is the distance from F *to the node of emergence*, n.

In the ordinary photographic lens, the internodal distance is small enough to be ignored for rough calculations. Obviously it is in the measurement of i that the need for caution arises. The telephoto lens is a special case in that the nodes are outside the lens mount at some distance in front of the lens (in the direction of the object). This explains the comparatively small camera extension required by these lenses in spite of their large focal lengths.

It can be shown easily that the nodes have the peculiar property that the lens may be rotated about an axis at right angles to the optic axis and containing the node without movement of the images of distant objects. This principle is used in the panoramic camera. It also provides a simple experimental method for determination of the position of the nodes.

INDEX

Printed in the United States
By Bookmasters